1948-1984

DEREK WALCOTT

COLLECTED POEMS

1948-1984

The Noonday Press

Farrar, Straus & Giroux

New York

Copyright © 1962, 1963, 1964, 1965, 1969, 1970, 1971, 1972, 1973,
1974, 1975, 1976, 1977, 1978, 1979, 1980, 1981, 1982, 1983, 1984,
and 1986 by Derek Walcott
All rights reserved
Printed in the United States of America
Published in Canada by HarperCollinsCanadaLtd
First edition, 1986
Fifth Printing, 1992

Certain of these poems first appeared in *The Agni Review, The American Poetry Review, Antaeus, Art and Man* (Trinidad), *The Beloit Poetry Journal, Between Worlds, Bim, Borestone Mountain Poetry Awards,* 1963, *Bostonia Magazine, The Bulletin* (Australia), *Caribbean Quarterly, Chicago Tribune, Columbia, Co-operation* (Canada), *Embers, Encounter, Evergreen Review, Fifteen Poems for William Shakespeare* (The Arts Council, England), *Guayana Festival Anthology, Harvard Advocate, Kenyon Review, London Magazine, Massachusetts Review, The Nation, New Letters, The New Statesman, New World Writing, The New York Review of Books, The New York Times Magazine, The New Yorker, Opus, The Pacific Quarterly Moana, The Paris Review, Persea, Review of English Literature, Savacou* (West Indies), *Spectator, Tamarack Review, Tapia* (Trinidad), *The Times* (London), and *Trinidad and Tobago Review*

Library of Congress Cataloging-in-Publication Data
Walcott, Derek.
Collected poems, 1948–1984.
Includes index.
I. Title.
PR9272.9.W3A17   1986   811   85-20688
ISBN 374-12626-7

For Alix Walcott

CONTENTS

From IN A GREEN NIGHT
Poems 1948–1960
[1962]

Prelude 3
As John to Patmos 5
A City's Death by Fire 6
The Harbour 7

From SELECTED POEMS
[1964]

Origins 11

From IN A GREEN NIGHT (1962)

A Far Cry from Africa 17
Ruins of a Great House 19
Tales of the Islands 22
Return to D'Ennery; Rain 28
Pocomania 31
Parang 33
Two Poems on the Passing of an Empire 35
Orient and Immortal Wheat 36
A Lesson for This Sunday 38
Bleecker Street, Summer 40
A Letter from Brooklyn 41

Brise Marine 43
A Sea-Chantey 44
The Polish Rider 47
The Banyan Tree, Old Year's Night 48
In a Green Night 50
Islands 52

From THE CASTAWAY AND OTHER POEMS [1965]

The Castaway 57
The Swamp 59
Tarpon 61
Missing the Sea 63
The Glory Trumpeter 64
A Map of Europe 66
Nights in the Gardens of Port of Spain 67
Crusoe's Island 68
Coral 73

From THE GULF [1970]

From THE CASTAWAY AND OTHER POEMS (1965)

The Flock 77
A Village Life 79
Goats and Monkeys 83
Laventille 85

Verandah 89
God Rest Ye Merry, Gentlemen 91
Crusoe's Journal 92
Lampfall 95
Codicil 97

From THE GULF AND OTHER POEMS (1969)

Mass Man 99
Exile 100
Homage to Edward Thomas 103
The Gulf 104
Elegy 109
Blues 111
Air 113
Guyana 115
Che 123
Negatives 124
Landfall, Grenada 125
Homecoming: Anse La Raye 127
Star 130
Cold Spring Harbor 131
Love in the Valley 133
Nearing Forty 136
The Walk 138

ANOTHER LIFE

[1973]

One / The Divided Child 143
Two / Homage to Gregorias 189
Three / A Simple Flame 223
Four / The Estranging Sea 259

From SEA GRAPES
[1976]

Sea Grapes 297
Sunday Lemons 298
New World 300
Adam's Song 302
Preparing for Exile 304
Names 305
Sainte Lucie 309
Volcano 324
Endings 326
The Fist 327
Love after Love 328
Dark August 329
Sea Canes 331
Midsummer, Tobago 333
Oddjob, a Bull Terrier 334
Winding Up 336
The Morning Moon 338
To Return to the Trees 339

From THE STAR-APPLE KINGDOM
[1979]

The Schooner *Flight* 345
Sabbaths, W.I. 362
The Sea Is History 364
Egypt, Tobago 368
The Saddhu of Couva 372
Forest of Europe 375
Koenig of the River 379
The Star-Apple Kingdom 383

From THE FORTUNATE TRAVELLER
[1981]

Old New England 399
Upstate 401
Piano Practice 403
North and South 405
Beachhead 410
Map of the New World 413
From This Far 414
Europa 418
The Man Who Loved Islands 420
Hurucan 423
Jean Rhys 427
The Liberator 430
The Spoiler's Return 432
The Hotel Normandie Pool 439
Early Pompeian 446
Easter 452
Wales 455
The Fortunate Traveller 456
The Season of Phantasmal Peace 464

From MIDSUMMER
[1984]

II / Companion in Rome, whom Rome makes as old as Rome 469
III / At the Queen's Park Hotel, with its white, high-ceilinged rooms 471
VI / Midsummer stretches beside me with its cat's yawn 472
VII / Our houses are one step from the gutter. Plastic curtains 474
XI / My double, tired of morning, closes the door 475
XIV / With the frenzy of an old snake shedding its skin 476

XV / I can sense it coming from far, too, Maman, the tide 477
XVIII / In the other 'eighties, a hundred midsummers gone 478
XIX / Gauguin 479
XX / Watteau 481
XXI / A long, white, summer cloud, like a cleared linen table 482
XXIII / With the stampeding hiss and scurry of green
       lemmings 483
XXV / The sun has fired my face to terra-cotta 484
XXVI / Before that thundercloud breaks from its hawsers 485
XXVII / Certain things here are quietly American— 486
XXVIII / Something primal in our spine makes the child
        swing 488
XXX / Gold dung and urinous straw from the horse garages 489
XXXIII / Those grooves in that forehead of sand-coloured
         flesh 490
XXXV / Mud. Clods. The sucking heel of the rain-flinger 491
XXXVI / The oak inns creak in their joints as light declines 492
XXXIX / The grey English road hissed emptily under
        the tires 493
XLI / The camps hold their distance—brown chestnuts and
      grey smoke 494
XLII / Chicago's avenues, as white as Poland 495
XLIII / Tropic Zone 496
XLIX / A wind-scraped headland, a sludgy, dishwater sea 503
L / I once gave my daughters, separately, two conch shells 504
LI / Since all of your work was really an effort to appease 505
LII / I heard them marching the leaf-wet roads of my head 506
LIII / There was one Syrian, with his bicycle, in our town 508
LIV / The midsummer sea, the hot pitch road, this grass,
      these shacks that made me 510

Books of Poetry by Derek Walcott 512
Index of Titles 513

From

IN A GREEN NIGHT:

POEMS 1948-1960

[1962]

Prelude

I, with legs crossed along the daylight, watch
The variegated fists of clouds that gather over
The uncouth features of this, my prone island.

Meanwhile the steamers which divide horizons prove
Us lost;
Found only
In tourist booklets, behind ardent binoculars;
Found in the blue reflection of eyes
That have known cities and think us here happy.

Time creeps over the patient who are too long patient,
So I, who have made one choice,
Discover that my boyhood has gone over.

And my life, too early of course for the profound cigarette,
The turned doorhandle, the knife turning
In the bowels of the hours, must not be made public
Until I have learnt to suffer
In accurate iambics.

I go, of course, through all the isolated acts,
Make a holiday of situations,
Straighten my tie and fix important jaws,

And note the living images
Of flesh that saunter through the eye.

Until from all I turn to think how,
In the middle of the journey through my life,
O how I came upon you, my
Reluctant leopard of the slow eyes.

—*1948*

As John to Patmos

As John to Patmos, among the rocks and the blue, live air, hounded
His heart to peace, as here surrounded
By the strewn-silver on waves, the wood's crude hair, the rounded
Breasts of the milky bays, palms, flocks, the green and dead

Leaves, the sun's brass coin on my cheek, where
Canoes brace the sun's strength, as John, in that bleak air,
So am I welcomed richer by these blue scapes, Greek there,
So I shall voyage no more from home; may I speak here.

This island is heaven—away from the dustblown blood of cities;
See the curve of bay, watch the straggling flower, pretty is
The wing'd sound of trees, the sparse-powdered sky, when lit is
The night. For beauty has surrounded
Its black children, and freed them of homeless ditties.

As John to Patmos, in each love-leaping air,
O slave, soldier, worker under red trees sleeping, hear
What I swear now, as John did:
To praise lovelong, the living and the brown dead.

A City's Death by Fire

After that hot gospeller had levelled all but the churched sky,
I wrote the tale by tallow of a city's death by fire;
Under a candle's eye, that smoked in tears, I
Wanted to tell, in more than wax, of faiths that were snapped
    like wire.
All day I walked abroad among the rubbled tales,
Shocked at each wall that stood on the street like a liar;
Loud was the bird-rocked sky, and all the clouds were bales
Torn open by looting, and white, in spite of the fire.
By the smoking sea, where Christ walked, I asked, why
Should a man wax tears, when his wooden world fails?
In town, leaves were paper, but the hills were a flock of faiths;
To a boy who walked all day, each leaf was a green breath
Rebuilding a love I thought was dead as nails,
Blessing the death and the baptism by fire.

The Harbour

The fishermen rowing homeward in the dusk
Do not consider the stillness through which they move,
So I, since feelings drown, should no more ask
For the safe twilight which your calm hands gave.
And the night, urger of old lies,
Winked at by stars that sentry the humped hills,
Should hear no secret faring-forth; time knows
That bitter and sly sea, and love raises walls.
Yet others who now watch my progress outward,
On a sea which is crueller than any word
Of love, may see in me the calm my passage makes,
Braving new water in an antique hoax;
And the secure from thinking may climb safe to liners
Hearing small rumours of paddlers drowned near stars.

From

SELECTED POEMS

[1964]

Origins

[*for Veronica Jenkin*]

narrow path of the surge in the blur of fables . . .
—CÉSAIRE

I

The flowering breaker detonates its surf.
White bees hiss in the coral skull.
Nameless I came among olives of algae,
Fœtus of plankton, I remember nothing.

Clouds, log of Colon,
I learnt your annals of ocean,
Of Hector, bridler of horses,
Achilles, Aeneas, Ulysses,
But "Of that fine race of people which came off the mainland
To greet Christobal as he rounded Icacos,"
Blank pages turn in the wind.
They possessed, by Bulbrook,
"No knowledge whatever of metals, not even of gold,
They recognized the seasons, the first risings of the Pleiades
By which signs they cultivate, assisted by magic . . .
Primitive minds cannot grasp infinity."

Nuages, nuages, in lazy volumes, rolled,
Swallowed in the surf of changing cumulus,
Their skulls of crackling shells crunched underfoot.
Now, when the mind would pierce infinity.
A gap in history closes, like a cloud.

    II

Memory in cerecloth uncoils its odour of rivers,
Of Egypt embalmed in an amber childhood.
In my warm, malarial bush-bath,
The wet leaves leeched to my flesh. An infant Moses,
I dreamed of dying, I saw
Paradise as columns of lilies and wheat-headed angels.

Between the Greek and African pantheon,
Lost animist, I rechristened trees:
Caduceus of Hermes: the constrictor round the mangrove.
Dorade, their golden, mythological dolphin,
Leapt, flaking light, as once for Arion,
For the broken archipelago of wave-browed gods.
Now, the sibyl I honour, mother of memory,
Bears in her black hand a white frangipani, with berries of blood,
She gibbers with the cries
Of the Guinean odyssey.

These islands have drifted from anchorage
Like gommiers loosened from Guinea,
Far from the childhood of rivers.

III

*O clear, brown tongue of the sun-warmed, sun-wooded Troumassee
of laundresses and old leaves, and winds that buried their old
songs in archives of bamboo and wild plantain, their white sails
bleached and beaten on dry stone, the handkerchiefs of adieux
and ba-bye! O sea, leaving your villages of cracked mud and*
    tin, your
*chorus of bearded corn in tragic fields, your children
like black rocks of petrified beginnings in whose potbellied
drought the hookworm boils, cherubim of glaucoma and gonorrhea!*
                     White cemeteries of shells beside the
*sea's cracked cobalt, poinsettia bleeding at your praying stations,
shadowy with croton and with "glory cedar,"*
      Whose gourds of cracked canoes hold the dead hopes
   of larvae,
*live middens heaped by the infecting river, fêtes of a childhood
brain sieved with sea-noise and river murmur,* ah, mon enfance.
      Ah, mon enfance! *Smothered in the cotton clouds of
illness, cocooned in sinuous odours of the censer, buried in bells,
bathed in the alcohol of lime and yellow flowers, voyaging like
Colon on starched, linen seas; that watched the river's*
    snakes writhe
*on the ceiling, that knew the forgotten taste of river water,
the odour of fresh bread and mother's skin, that knew its own skin
slowly (amber, then excrement, then bronze), that feared the*
    Ibo fifes
and drums of Christmas, the broken egg in which it sailed
   at Easter,
festivals, processions, voyages of the grave, and
   the odour of rivers in unopened cupboards.

IV

Trace of our exodus across its desert
Erased by the salt winds.

The snake spirit dies, writhing horizon.
Beetles lift the dead elephant into the jaws of the forest.
Death of old gods in the ashes of their eyes.
The plunging throats of porpoises simulating, O sea,
The retching hulks of caravels stitching two worlds,
Like the whirr of my mother's machine in a Sabbath bedroom,
Like needles of cicadas stitching the afternoon's shroud.
Death of old gods in the river snakes dried from the ceiling,
Jahveh and Zeus rise from the foam's beard at daybreak.

The mind, among sea-wrack, sees its mythopoeic coast,
Seeks, like the polyp, to take root in itself.
Here, in the rattle of receding shoal,
Among these shallows I seek my own name and a man.
As the crab's claws move backwards through the surf,
Blind memory grips the putrefying flesh.

V

Was it not then we asked for a new song,
As Colon's vision gripped the berried branch?
For the names of bees in the surf of white frangipani,
With hard teeth breaking the bitter almonds of consonants,
Shaping new labials to the curl of the wave,
Christening the pomegranate with a careful tongue,
Pommes de Cythère, bitter Cytherean apple.
And God's eye glazed by an indifferent blue.

VI

We have learnt their alphabet of alkali and aloe, on seeds of islands dispersed by the winds. We have washed out with salt the sweet, faded savour of rivers, and in the honeycombs of skulls the bees built a new song. And we have eaten of their bitter olive. But now, twin soul, spirit of river, spirit of sea, turn from the long, interior rivers, their somnolence, brown studies, their long colonial languor, their old Egyptian sickness, their imitation of tea colour, their tongues that lick the feet of bwana and sahib, their rage for funeral pyres of children's flesh, their sinuousness that shaped the original snake. The surf has razed that
    memory from
our speech, and
       a single raindrop irrigates the tongue.

VII

The sea waits for him, like Penelope's spindle,
Ravelling, unravelling its foam,
Whose eyes bring the rain from far countries, the salt rain
That hazes horizons and races,
Who, crouched by our beach fires, his face cracked by deserts,
Remembering monarchs, asks us for water
Fetched in the fragment of an earthen cruse,
And extinguishes Troy in a hissing of ashes,
In a rising of cloud.

Clouds, vigorous exhalations of wet earth,
In men and in beasts the nostrils exulting in rain scent,
Uncoiling like mist, the wound of the jungle,
We praise those whose back on hillsides buckles on the wind

To sow the grain of Guinea in the mouths of the dead,
Who, hurling their bone-needled nets over the cave mouth,
Harvest ancestral voices from its surf,
Who, lacking knowledge of metals, primarily of gold,
Still gather the coinage of cowries, simple numismatists,
Who kneel in the open sarcophagi of cocoa
To hallow the excrement of our martyrdom and fear,
Whose sweat, touching earth, multiplies in crystals of sugar
Those who conceive the birth of white cities in a raindrop
And the annihilation of races in the prism of the dew.

From IN A GREEN NIGHT
[1962]

A Far Cry from Africa

A wind is ruffling the tawny pelt
Of Africa. Kikuyu, quick as flies,
Batten upon the bloodstreams of the veldt.
Corpses are scattered through a paradise.
Only the worm, colonel of carrion, cries:
"Waste no compassion on these separate dead!"
Statistics justify and scholars seize
The salients of colonial policy.
What is that to the white child hacked in bed?
To savages, expendable as Jews?

Threshed out by beaters, the long rushes break
In a white dust of ibises whose cries
Have wheeled since civilization's dawn
From the parched river or beast-teeming plain.
The violence of beast on beast is read
As natural law, but upright man
Seeks his divinity by inflicting pain.
Delirious as these worried beasts, his wars
Dance to the tightened carcass of a drum,
While he calls courage still that native dread
Of the white peace contracted by the dead.

Again brutish necessity wipes its hands
Upon the napkin of a dirty cause, again

A waste of our compassion, as with Spain,
The gorilla wrestles with the superman.
I who am poisoned with the blood of both,
Where shall I turn, divided to the vein?
I who have cursed
The drunken officer of British rule, how choose
Between this Africa and the English tongue I love?
Betray them both, or give back what they give?
How can I face such slaughter and be cool?
How can I turn from Africa and live?

Ruins of a Great House

> *though our longest sun sets at right declensions and*
> *makes but winter arches, it cannot be long before we*
> *lie down in darkness, and have our light in ashes . . .*
> —BROWNE, *Urn Burial*

Stones only, the disjecta membra of this Great House,
Whose moth-like girls are mixed with candledust,
Remain to file the lizard's dragonish claws.
The mouths of those gate cherubs shriek with stain;
Axle and coach wheel silted under the muck
Of cattle droppings.
                    Three crows flap for the trees
And settle, creaking the eucalyptus boughs.
A smell of dead limes quickens in the nose
The leprosy of empire.
              "Farewell, green fields,
                  Farewell, ye happy groves!"
Marble like Greece, like Faulkner's South in stone,
Deciduous beauty prospered and is gone,
But where the lawn breaks in a rash of trees
A spade below dead leaves will ring the bone
Of some dead animal or human thing
Fallen from evil days, from evil times.

It seems that the original crops were limes
Grown in the silt that clogs the river's skirt;
The imperious rakes are gone, their bright girls gone,
The river flows, obliterating hurt.
I climbed a wall with the grille ironwork
Of exiled craftsmen protecting that great house
From guilt, perhaps, but not from the worm's rent
Nor from the padded cavalry of the mouse.
And when a wind shook in the limes I heard
What Kipling heard, the death of a great empire, the abuse
Of ignorance by Bible and by sword.

A green lawn, broken by low walls of stone,
Dipped to the rivulet, and pacing, I thought next
Of men like Hawkins, Walter Raleigh, Drake,
Ancestral murderers and poets, more perplexed
In memory now by every ulcerous crime.
The world's green age then was a rotting lime
Whose stench became the charnel galleon's text.
The rot remains with us, the men are gone.
But, as dead ash is lifted in a wind
That fans the blackening ember of the mind,
My eyes burned from the ashen prose of Donne.

Ablaze with rage I thought,
Some slave is rotting in this manorial lake,
But still the coal of my compassion fought
That Albion too was once
A colony like ours, "part of the continent, piece of the main,"
Nook-shotten, rook o'erblown, deranged
By foaming channels and the vain expense
Of bitter faction.

All in compassion ends
So differently from what the heart arranged:
"as well as if a manor of thy friend's . . ."

Tales of the Islands

Chapter I / La rivière dorée . . .

The marl white road, the Dorée rushing cool
Through gorges of green cedars, like the sound
Of infant voices from the Mission School,
Like leaves like dim seas in the mind; *ici*, Choiseul.
The stone cathedral echoes like a well,
Or as a sunken sea-cave, carved, in sand.
Touring its Via Dolorosa I tried to keep
That chill flesh from my memory when I found
A Sancta Teresa in her nest of light;
The skirts of fluttered bronze, the uplifted hand,
The cherub, shaft upraised, parting her breast.
Teach our philosophy the strength to reach
Above the navel; black bodies, wet with light,
Rolled in the spray as I strolled up the beach.

Chapter II / "Qu'un sang impur . . ."

Cosimo de Chrétien controlled a boardinghouse.
His maman managed him. No. 13.
Rue St. Louis. It had a court, with rails,
A perroquet, a curio shop where you
Saw black dolls and an old French barquentine
Anchored in glass. Upstairs, the family sword,

The rusting icon of a withered race,
Like the first angel's kept its pride of place,
Reminding the bald count to keep his word
Never to bring the lineage to disgrace.
Devouring Time, which blunts the Lion's claws,
Kept Cosimo, count of curios, fairly chaste,
For Mama's sake, for hair oil, and for whist;
Peering from balconies for his tragic twist.

Chapter III / La belle qui fut . . .

Miss Rossignol lived in the lazaretto
For Roman Catholic crones; she had white skin,
And underneath it, fine, old-fashioned bones;
She flew like bats to vespers every twilight,
The living Magdalen of Donatello;
And tipsy as a bottle when she stalked
On stilted legs to fetch the morning milk,
In a black shawl harnessed by rusty brooches.
My mother warned us how that flesh knew silk
Coursing a green estate in gilded coaches.
While Miss Rossignol, in the cathedral loft,
Sang to her one dead child, a tattered saint
Whose pride had paupered beauty to this witch
Who was so fine once, whose hands were so soft.

Chapter IV / "Dance of Death"

Outside I said, "He's a damned epileptic,
Your boy, El Greco! Goya, he don't lie."
Doc laughed: "Let's join the real epileptics."
Two of the girls looked good. The Indian said

That rain affects the trade. In the queer light
We all looked green. The beer and all looked green.
One draped an arm around me like a wreath.
The next talked politics. "Our mother earth,"
I said. "The great republic in whose womb
The dead outvote the quick." "Y'all too obscene,"
The Indian laughed. "Y'all college boys ain't worth
The trouble." We entered the bare room.
In the rain, walking home, was worried, but Doc said:
"Don't worry, kid, the wages of sin is birth."

Chapter V / "moeurs anciennes"

The fête took place one morning in the heights
For the approval of some anthropologist.
The priests objected to such savage rites
In a Catholic country; but there was a twist
As one of the fathers was himself a student
Of black customs; it was quite ironic.
They lead sheep to the rivulet with a drum,
Dancing with absolutely natural grace
Remembered from the dark past whence we come.
The whole thing was more like a bloody picnic.
Bottles of white rum and a brawling booth.
They tie the lamb up, then chop off the head,
And ritualists take turns drinking the blood.
Great stuff, old boy; sacrifice, moments of truth.

Chapter VI

Poopa, da' was a fête! I mean it had
Free rum free whisky and some fellars beating

Pan from one of them band in Trinidad,
And everywhere you turn was people eating
And drinking and don't name me but I think
They catch his wife with two tests up the beach
While he drunk quoting Shelley with "Each
Generation has its angst, but we has none"
And wouldn't let a comma in edgewise.
(Black writer chap, one of them Oxbridge guys.)
And it was round this part once that the heart
Of a young child was torn from it alive
By two practitioners of native art,
But that was long before this jump and jive.

    Chapter VII / Lotus eater ...

"Maingot," the fisherman called that pool blocked by
Increasing filth that piled between ocean
And jungle, with a sighing grove
Of dry bamboo, its roots freckled with light
Like feathers fallen from a migratory sky.
Beyond that, the village. Through urine-stunted trees
A mud path wriggled like a snake in flight.
Franklin gripped the bridge stanchions with a hand
Trembling from fever. Each spring, memories
Of his own country where he could not die
Assaulted him. He watched the malarial light
Shiver the canes. In the tea-coloured pool, tadpoles
Seemed happy in their element. Poor, black souls.
He shook himself. Must breed, drink, rot with motion.

Chapter VIII

In the Hotel Miranda, 10 Grass St., who fought
The Falangists *en la guerra civil*, at the hour
Of bleeding light and beads of crimson dew,
This exile, with the wry face of a Jew,
Lets dust powder his pamphlets; crook't
Fingers clutch a journal to his shirt.
The eye is glacial; mountainous, the hook'd
Nose down which an ant, caballo, rides. Besides,
As pious fleas explore a seam of dirt,
The sunwashed body, past the age of sweat,
Sprawls like a hero, curiously inert.
Near him a dish of olives has turned sour.
Above the children's street cries, a girl plays
A marching song not often sung these days.

Chapter IX/"Le loupgarou"

A curious tale that threaded through the town
Through greying women sewing under eaves,
Was how his greed had brought old Le Brun down,
Greeted by slowly shutting jalousies
When he approached them in white linen suit,
Pink glasses, cork hat, and tap-tapping cane,
A dying man licensed to sell sick fruit,
Ruined by fiends with whom he'd made a bargain.
It seems one night, these Christian witches said,
He changed himself to an Alsatian hound,
A slavering lycanthrope hot on a scent,
But his own watchman dealt the thing a wound.

It howled and lugged its entrails, trailing wet
With blood, back to its doorstep, almost dead.

> *Chapter X / "Adieu foulard . . ."*

I watched the island narrowing the fine
Writing of foam around the precipices, then
The roads as small and casual as twine
Thrown on its mountains; I watched till the plane
Turned to the final north and turned above
The open channel with the grey sea between
The fishermen's islets until all that I love
Folded in cloud; I watched the shallow green
That broke in places where there would be reef,
The silver glinting on the fuselage, each mile
Dividing us and all fidelity strained
Till space would snap it. Then, after a while
I thought of nothing; nothing, I prayed, would change;
When we set down at Seawell it had rained.

Return to D'Ennery; Rain

Imprisoned in these wires of rain, I watch
This village stricken with a single street,
Each weathered shack leans on a wooden crutch,
Contented as a cripple with defeat.
Five years ago even poverty seemed sweet,
So azure and indifferent was this air,
So murmurous of oblivion the sea,
That any human action seemed a waste,
The place seemed born for being buried there.
                              The surf explodes
In scissor-birds hunting the usual fish,
The rain is muddying unpaved inland roads,
So personal grief melts in the general wish.

The hospital is quiet in the rain.
A naked boy drives pigs into the bush.
The coast shudders with every surge. The beach
Admits a beaten heron. Filth and foam.
There in a belt of emerald light, a sail
Plunges and lifts between the crests of reef,
The hills are smoking in the vaporous light,
The rain seeps slowly to the core of grief.
It could not change its sorrows and be home.

It cannot change, though you become a man
Who would exchange compassion for a drink,
Now you are brought to where manhood began
Its separation from "the wounds that make you think."
And as this rain puddles the sand, it sinks
Old sorrows in the gutter of the mind;
Where is that passionate hatred that would help
The black, the despairing, the poor, by speech alone?
The fury shakes like wet leaves in the wind,
The rain beats on a brain hardened to stone.

For there is a time in the tide of the heart, when
Arrived at its anchor of suffering, a grave
Or a bed, despairing in action, we ask,
O God, where is our home? For no one will save
The world from itself, though he walk among men,
On such shores where the foam
Murmurs oblivion of action, who raise
No cry like herons stoned by the rain.

The passionate exiles believe it, but the heart
Is circled by sorrows, by its horror
And bitter devotion to home.
The romantic nonsense ends at the bowsprit, shearing
But never arriving beyond the reef-shore foam,
Or the rain cuts us off from heaven's hearing.

Why blame the faith you have lost? Heaven remains
Where it is, in the hearts of these people,
In the womb of their church, though the rain's
Shroud is drawn across its steeple.

You are less than they are, for your truth
Consists of a general passion, a personal need,
Like that ribbed wreck, abandoned since your youth,
Washed over by the sour waves of greed.

The white rain draws its net along the coast,
A weak sun streaks the villages and beaches
And roads where laughing labourers come from shelter,
On heights where charcoal-burners heap their days.
Yet in you it still seeps, blurring each boast
Your craft has made, obscuring words and features,
Nor have you changed from all of the known ways
To leave the mind's dark cave, the most
Accursed of God's self-pitying creatures.

Pocomania

De shepherd shrieves in Egyptian light,
The Abyssinian sweat has poured
From armpits and the graves of sight,
The black sheep of their blacker Lord.

De sisters shout and lift the floods
Of skirts where bark 'n' balm take root,
De bredren rattle withered gourds
Whose seeds are the forbidden fruit.

Remorse of poverty, love of God
Leap as one fire; prepare the feast,
Limp now is each divining rod,
Forgotten love, the double beast.

Above the banner and the crowd
The Lamb bleeds on the Coptic cross,
De Judah Lion roars to shroud
The sexual fires of Pentecost.

In jubilation of the Host,
The goatskin greets the bamboo fife,
Have mercy on these furious lost
Whose life is praising death in life.

Now the blind beast butts on the wall,
Bodily delirium is death,
Now the worm curls upright to crawl
Between the crevices of breath.

Lower the wick, and fold the eye!
Anoint the shrivelled limb with oil!
The waters of the moon are dry,
Derision of the body, toil.

Till Armageddon stains the fields,
And Babylon is yonder green,
Till the dirt-holy roller feels
The obscene breeding the unseen.

Till those black forms be angels white,
And Zion fills each eye.
High overhead the crow of night
Patrols eternity.

Parang

Man, I suck me tooth when I hear
How dem croptime fiddlers lie,
And de wailing, kiss-me-arse flutes
That bring water to me eye!
Oh, when I t'ink how from young
I wasted time at de fêtes,
I could bawl in a red-eyed rage
For desire turned to regret,
Not knowing the truth that I sang
At parang and *la commette*.
Boy, every damned tune them tune
Of love that go last forever
Is the wax and the wane of the moon
Since Adam catch body-fever.

I old, so the young crop won't
Have these claws to reap their waist,
But I know "do more" from "don't"
Since the grave cry out "Make haste!"
This banjo world have one string
And all man does dance to that tune:
That love is a place in the bush
With music grieving from far,
As you look past her shoulder and see
Like her one tear afterwards

The falling of a fixed star.
Young men does bring love to disgrace
With remorseful, regretful words,
When flesh upon flesh was the tune
Since the first cloud raise up to disclose
The breast of the naked moon.

Two Poems on the Passing of an Empire

I

A heron flies across the morning marsh and brakes
its teetering wings to decorate a stump
                              (thank God
that from this act the landscape is complete
and time and motion at a period
as such an emblem led Rome's trampling feet,
pursued by late proconsuls bearing law)
and underline this quiet with a caw.

II

In the small coffin of his house, the pensioner,
A veteran of the African campaign,
Bends, as if threading an eternal needle;
One-eyed as any grave, his skull, cropped wool,
Or lifts his desert squint to hear
The children singing, "Rule, Britannia, rule,"
As if they needed practise to play dead.
Boys will still pour their blood out for a sieve
Despite his balsam eye and doddering jaw;
And if one eye should weep, would they believe
In such a poor flag as an empty sleeve?

Orient and Immortal Wheat

> *The corn was orient and immortal wheat, which never should be reaped, nor was ever sown. I thought it had stood from everlasting to everlasting.*
> —TRAHERNE, *Centuries of Meditations*

Nature seemed monstrous to his thirteen years.
Prone to malaria, sweating inherent sin,
Absolved in Limacol and evening prayers,
The prodigy, dusk rouging his peaked face,
Studied the swallows stitch the opposing eaves
In repetitions of the fall from grace.
And as a gilding silence flushed the leaves,
Hills, roofs, and yards with his own temperature,
He wept again, though why, he was unsure,
At dazzling visions of reflected tin.
So heaven is revealed to fevered eyes,
So is sin born, and innocence made wise,
By intimations of hot galvanize.

This was the fever called original sin,
Such anthropomorphic love illumines hell,
A charge brought to his Heavenly Father's face
That wept for bat-voiced orphans in the streets
And cripples limping homeward in weak light,
When the lamplighter, his head swung by its hair,

Meant the dread footfall lumping up the stair:
Maman with soup, perhaps; or it could well
Be Chaos, genderer of Earth, called Night.

A Lesson for This Sunday

The growing idleness of summer grass
With its frail kites of furious butterflies
Requests the lemonade of simple praise
In scansion gentler than my hammock swings
And rituals no more upsetting than a
Black maid shaking linen as she sings
The plain notes of some Protestant hosanna—
Since I lie idling from the thought in things—

Or so they should, until I hear the cries
Of two small children hunting yellow wings,
Who break my Sabbath with the thought of sin.
Brother and sister, with a common pin,
Frowning like serious lepidopterists.
The little surgeon pierces the thin eyes.
Crouched on plump haunches, as a mantis prays
She shrieks to eviscerate its abdomen.
The lesson is the same. The maid removes
Both prodigies from their interest in science.
The girl, in lemon frock, begins to scream
As the maimed, teetering thing attempts its flight.
She is herself a thing of summery light,
Frail as a flower in this blue August air,
Not marked for some late grief that cannot speak.

The mind swings inward on itself in fear
Swayed towards nausea from each normal sign.
Heredity of cruelty everywhere,
And everywhere the frocks of summer torn,
The long look back to see where choice is born,
As summer grass sways to the scythe's design.

Bleecker Street, Summer

Summer for prose and lemons, for nakedness and languor,
for the eternal idleness of the imagined return,
for rare flutes and bare feet, and the August bedroom
of tangled sheets and the Sunday salt, ah violin!

When I press summer dusks together, it is
a month of street accordions and sprinklers
laying the dust, small shadows running from me.

It is music opening and closing, *Italia mia*, on Bleecker,
ciao, Antonio, and the water-cries of children
tearing the rose-coloured sky in streams of paper;
it is dusk in the nostrils and the smell of water
down littered streets that lead you to no water,
and gathering islands and lemons in the mind.

There is the Hudson, like the sea aflame.
I would undress you in the summer heat,
and laugh and dry your damp flesh if you came.

A Letter from Brooklyn

An old lady writes me in a spidery style,
Each character trembling, and I see a veined hand
Pellucid as paper, travelling on a skein
Of such frail thoughts its thread is often broken;
Or else the filament from which a phrase is hung
Dims to my sense, but caught, it shines like steel,
As touch a line and the whole web will feel.
She describes my father, yet I forget her face
More easily than my father's yearly dying;
Of her I remember small, buttoned boots and the place
She kept in our wooden church on those Sundays
Whenever her strength allowed;
Grey-haired, thin-voiced, perpetually bowed.

"I am Mable Rawlins," she writes, "and know both your parents";
He is dead, Miss Rawlins, but God bless your tense:
"Your father was a dutiful, honest,
Faithful, and useful person."
For such plain praise what fame is recompense?
"A horn-painter, he painted delicately on horn,
He used to sit around the table and paint pictures."
The peace of God needs nothing to adorn
It, nor glory nor ambition.
"He is twenty-eight years buried," she writes, "he was called home,
And is, I am sure, doing greater work."

The strength of one frail hand in a dim room
Somewhere in Brooklyn, patient and assured,
Restores my sacred duty to the Word.
"Home, home," she can write, with such short time to live,
Alone as she spins the blessings of her years;
Not withered of beauty if she can bring such tears,
Nor withdrawn from the world that breaks its lovers so;
Heaven is to her the place where painters go,
All who bring beauty on frail shell or horn,
There was all made, thence their *lux-mundi* drawn,
Drawn, drawn, till the thread is resilient steel,
Lost though it seems in darkening periods,
And there they return to do work that is God's.

So this old lady writes, and again I believe.
I believe it all, and for no man's death I grieve.

Brise Marine

K with quick laughter, honey skin and hair,
and always money. In what beach shade, what year
has she so scented with her gentleness
I cannot watch bright water but think of her
and that fine morning when she sang O rare
Ben's lyric of "the bag o' the bee"
and "the nard in the fire"
                          "nard in the fire"
against the salty music of the sea
the fresh breeze tangling each honey tress
                         and what year was the fire?
Girls' faces dim with time, Andreuille all gold ...
Sunday. The grass peeps through the breaking pier.
Tables in the trees, like entering Renoir.
Maintenant je n'ai plus ni fortune, ni pouvoir ...
But when the light was setting through thin hair,
Holding whose hand by what trees, what old wall.

Two honest women, Christ, where are they gone?
Out of that wonder, what do I recall?
The darkness closing round a fisherman's oar.
The sound of water gnawing at bright stone.

A Sea-Chantey

Là, tout n'est qu'ordre et beauté,
Luxe, calme, et volupté.
—BAUDELAIRE

Anguilla, Adina,
Antigua, Cannelles,
Andreuille, all the *l*'s,
Voyelles, of the liquid Antilles,
The names tremble like needles
Of anchored frigates,
Yachts tranquil as lilies,
In ports of calm coral,
The lithe, ebony hulls
Of strait-stitching schooners,
The needles of their masts
That thread archipelagoes
Refracted embroidery
In feverish waters
Of the seafarer's islands,
Their shorn, leaning palms,
Shaft of Odysseus,
Cyclopic volcanoes,
Creak their own histories,
In the peace of green anchorage;
Flight, and Phyllis,

Returned from the Grenadines,
Names entered this Sabbath,
In the port clerk's register;
Their baptismal names,
The sea's liquid letters,
Repos donnez à cils . . .
And their blazing cargoes
Of charcoal and oranges;
Quiet, the fury of their ropes.
Daybreak is breaking
On the green chrome water,
The white herons of yachts
Are at Sabbath communion,
The histories of schooners
Are murmured in coral,
Their cargoes of sponges
On sandspits of islets,
Barques white as white salt
Of acrid St. Maarten,
Hulls crusted with barnacles,
Holds foul with great turtles,
Whose ship-boys have seen
The blue heave of Leviathan,
A seafaring, Christian,
And intrepid people.

Now an apprentice washes his cheeks
With salt water and sunlight.

In the middle of the harbour
A fish breaks the Sabbath
With a silvery leap.

The scales fall from him
In a tinkle of church bells;
The town streets are orange
With the week-ripened sunlight,
Balanced on the bowsprit
A young sailor is playing
His grandfather's chantey
On a trembling mouth organ;
The music curls, dwindling
Like smoke from blue galleys,
To dissolve near the mountains.
The music uncurls with
The soft vowels of inlets,
The christening of vessels,
The titles of portages,
The colours of sea grapes,
The tartness of sea-almonds,
The alphabet of church bells,
The peace of white horses,
The pastures of ports,
The litany of islands,
The rosary of archipelagoes,
Anguilla, Antigua,
Virgin of Guadeloupe,
And stone-white Grenada
Of sunlight and pigeons,
The amen of calm waters,
The amen of calm waters,
The amen of calm waters.

The Polish Rider

The grey horse, Death, in profile bears the young Titus
To dark woods by the dying coal of day;
The father with worn vision portrays the son
Like Dürer's knight astride a Rosinante;
The horse disturbs more than the youth delights us.
The warrior turns his sure gaze for a second,
Assurance looks its father in the eye,
The inherited, bony hack heads accurately
Towards the symbolic forests that have beckoned
Such knights, squired by the scyther, where to lie.
But skill dispassionately praises the rider,
Despair details the grey, cadaverous steed,
The immortal image holds its murderer
In a clear gaze for the next age to read.

The Banyan Tree, Old Year's Night

I

In the damp park, no larger than a stamp,
The rainbow bulbs of the year's end are looped
To link the withered fountain, and each lamp
Flickers like echoes where small savages whooped.

The square was this town's centre, but its spokes
Burn like a petered pinwheel of dead streets,
Turning in mind the squibs of boyish jokes,
Candy-striped innocents and sticky sweets

Fading in lemon light, as ribbons fade;
Bring back the pumping major and the snails
Of tubas marching as the brass band played
For children punished in their window gaols,

And gusts of tumbling papers, babies, kites
Blown round the kiosk band rails in the wind;
But now these ghosts like wan bulbs show the whites
Of vanished eyes, and absence crowds the mind.

Soaring from littered roots, blackened with rain,
With inaccessible arms the banyan tree
Heaves in the year's last drizzle to explain
What age could not, responsibility.

II

At this town's rotting edges foul canals
Race with assurance when bad weather pours
White rain and wind by which the paper sails
Of crouched black children steer for little tours

Till the silt clogs them on the farther bank;
And the barques tilt, sunk in short voyages.
Yet, as they dare each season, so I thank
What wind compelled my flight, whatever rages

Urged my impossible exile; so with this park
I study now, as exiles stamp from home,
Fearing those bulbs will hiss out in the dark,
The mind be swept of truths as by a broom.

Even on silvery days, that classic fount
Being withered to the root, its throat as hoarse
As the last nurse's cry, could not surmount
My growing fear with clarity from a source

No parent knew. Or did we march
To the brass tunes of truth? Did I divine
Some secret in the fountain's failing arch,
And was that infant melancholy mine?

If it were so, it still remains, its sources
Blank as the rain on the deserted mind,
Dumb as the ancient Indian tree that forces
Its grieving arms to keep the homeless wind.

In a Green Night

The orange tree, in various light,
Proclaims perfected fables now
That her last season's summer height
Bends from each overburdened bough.

She has her winters and her spring,
Her moult of leaves, which in their fall
Reveal, as with each living thing,
Zones truer than the tropical.

For if by night each golden sun
Burns in a comfortable creed,
By noon harsh fires have begun
To quail those splendours which they feed.

Or mixtures of the dew and dust
That early shone her orbs of brass,
Mottle her splendours with the rust
She sought all summer to surpass.

By such strange, cyclic chemistry
That dooms and glories her at once
As green yet aging orange tree,
The mind enspheres all circumstance.

No Florida loud with citron leaves
With crystal falls to heal this age
Shall calm the darkening fear that grieves
The loss of visionary rage.

Or if Time's fires seem to blight
The nature ripening into art,
Not the fierce noon or lampless night
Can quail the comprehending heart.

The orange tree, in various light,
Proclaims that fable perfect now
That her last season's summer height
Bends from each overburdened bough.

Islands

[*for Margaret*]

Merely to name them is the prose
Of diarists, to make you a name
For readers who like travellers praise
Their beds and beaches as the same;
But islands can only exist
If we have loved in them. I seek,
As climate seeks its style, to write
Verse crisp as sand, clear as sunlight,
Cold as the curled wave, ordinary
As a tumbler of island water;
Yet, like a diarist, thereafter
I savour their salt-haunted rooms
(Your body stirring the creased sea
Of crumpled sheets), whose mirrors lose
Our huddled, sleeping images,
Like words which love had hoped to use
Erased with the surf's pages.

So, like a diarist in sand,
I mark the peace with which you graced
Particular islands, descending
A narrow stair to light the lamps
Against the night surf's noises, shielding

A leaping mantle with one hand,
Or simply scaling fish for supper,
Onions, jack-fish, bread, red snapper;
And on each kiss the harsh sea-taste,
And how by moonlight you were made
To study most the surf's unyielding
Patience though it seems a waste.

From

THE CASTAWAY

AND OTHER POEMS

[1965]

The Castaway

The starved eye devours the seascape for the morsel
Of a sail.

The horizon threads it infinitely.

Action breeds frenzy. I lie,
Sailing the ribbed shadow of a palm,
Afraid lest my own footprints multiply.

Blowing sand, thin as smoke,
Bored, shifts its dunes.
The surf tires of its castles like a child.

The salt green vine with yellow trumpet-flower,
A net, inches across nothing.
Nothing: the rage with which the sandfly's head is filled.

Pleasures of an old man:
Morning: contemplative evacuation, considering
The dried leaf, nature's plan.

In the sun, the dog's feces
Crusts, whitens like coral.
We end in earth, from earth began.
In our own entrails, genesis.

If I listen I can hear the polyp build,
The silence thwanged by two waves of the sea.
Cracking a sea-louse, I make thunder split.

Godlike, annihilating godhead, art
And self, I abandon
Dead metaphors: the almond's leaf-like heart,

The ripe brain rotting like a yellow nut
Hatching
Its babel of sea-lice, sandfly, and maggot,

That green wine bottle's gospel choked with sand,
Labelled, a wrecked ship,
Clenched sea-wood nailed and white as a man's hand.

The Swamp

Gnawing the highway's edges, its black mouth
Hums quietly: "Home, come home..."

Behind its viscous breath the very word "growth"
Grows fungi, rot;
White mottling its root.

More dreaded
Than canebrake, quarry, or sun-shocked gully bed,
Its horrors held Hemingway's hero rooted
To sure, clear shallows.

It begins nothing. Limbo of cracker convicts, Negroes.
Its black mood
Each sunset takes a smear of your life's blood.

Fearful, original sinuosities! Each mangrove sapling
Serpentlike, its roots obscene
As a six-fingered hand,

Conceals within its clutch the mossbacked toad,
Toadstools, the potent ginger-lily,
Petals of blood,

The speckled vulva of the tiger-orchid;
Outlandish phalloi
Haunting the travellers of its one road.

Deep, deeper than sleep
Like death,
Too rich in its decrescence, too close of breath,

In the fast-filling night, note
How the last bird drinks darkness with its throat,
How the wild saplings slip

Backward to darkness, go black
With widening amnesia, take the edge
Of nothing to them slowly, merge

Limb, tongue, and sinew into a knot
Like chaos, like the road
Ahead.

Tarpon

At Cedros, thudding the dead sand
in spasms, the tarpon
gaped with a gold eye, drowned
thickly, thrashing with brute pain
this sea I breathe.
Stilled, its bulk,
screwed to the eye's lens, slowly
sought design. It dried like silk,
leisurely, altered to lead.
The belly, leprous, silver, bulged
like a cold chancre for the blade.
Suddenly it shuddered in immense
doubt, but the old jaw, gibbering, divulged
nothing but some new filaments
of blood. For every bloody stroke
with which a frenzied fisherman struck
its head my young son shook his head.
Could I have called out not to look
simply at the one world we shared?
Dead, and examined in detail,
a tarpon's bulk grows beautiful.

Bronze, with a brass-green mould, the scales
age like a corselet of coins,
a net of tarnished silver joins

the back's deep-sea blue to the tail's
wedged, tapering Y.
Set in a stone, triangular skull,
ringing with gold, the open eye
is simply, tiringly there.
A shape so simple, like a cross,
a child could draw it in the air.
A tarpon's scale, its skin's flake
washed at the sea's edge and held
against the light, looks just like what
the grinning fisherman said it would:
dense as frost glass but delicate,
etched by a diamond, it showed
a child's drawing of a ship,
the sails' twin triangles, a mast.

Can such complexity of shape,
such bulk, terror, and fury fit
in a design so innocent,
that through opaque, phantasmal mist,
moving, but motionlessly, it
sails where imagination sent?

Missing the Sea

Something removed roars in the ears of this house,
Hangs its drapes windless, stuns mirrors
Till reflections lack substance.

Some sound like the gnashing of windmills ground
To a dead halt;
A deafening absence, a blow.

It hoops this valley, weighs this mountain,
Estranges gesture, pushes this pencil
Through a thick nothing now,

Freights cupboards with silence, folds sour laundry
Like the clothes of the dead left exactly
As the dead behaved by the beloved,

Incredulous, expecting occupancy.

The Glory Trumpeter

Old Eddie's face, wrinkled with river lights,
Looked like a Mississippi man's. The eyes,
Derisive and avuncular at once,
Swivelling, fixed me. They'd seen
Too many wakes, too many cathouse nights.
The bony, idle fingers on the valves
Of his knee-cradled horn could tear
Through "Georgia on My Mind" or "Jesus Saves"
With the same fury of indifference
If what propelled such frenzy was despair.

Now, as the eyes sealed in the ashen flesh,
And Eddie, like a deacon at his prayer,
Rose, tilting the bright horn, I saw a flash
Of gulls and pigeons from the dunes of coal
Near my grandmother's barracks on the wharves,
I saw the sallow faces of those men
Who sighed as if they spoke into their graves
About the Negro in America. That was when
The Sunday comics, sprawled out on her floor,
Sent from the States, had a particular odour;
Dry smell of money mingled with man's sweat.

And yet, if Eddie's features held our fate,
Secure in childhood I did not know then

A jesus-ragtime or gut-bucket blues
To the bowed heads of lean, compliant men
Back from the States in their funereal serge,
Black, rusty homburgs and limp waiters' ties,
Slow, honey accents and lard-coloured eyes,
Was Joshua's ram's horn wailing for the Jews
Of patient bitterness or bitter siege.

Now it was that, as Eddie turned his back
On our young crowd out fêting, swilling liquor,
And blew, eyes closed, one foot up, out to sea,
His horn aimed at those cities of the Gulf,
Mobile and Galveston, and sweetly meted
Their horn of plenty through his bitter cup,
In lonely exaltation blaming me
For all whom race and exile have defeated,
For my own uncle in America,
That living there I never could look up.

A Map of Europe

Like Leonardo's idea
Where landscapes open on a waterdrop
Or dragons crouch in stains,
My flaking wall, in the bright air,
Maps Europe with its veins.

On its limned window ledge
A beer can's gilded rim gleams like
Evening along a Canaletto lake,
Or like that rocky hermitage
Where, in his cell of light, haggard Jerome
Prays that His kingdom come
To the far city.

The light creates its stillness. In its ring
Everything IS. A cracked coffee cup,
A broken loaf, a dented urn become
Themselves, as in Chardin,
Or in beer-bright Vermeer,
Not objects of our pity.

In it is no *lacrimae rerum*,
No art. Only the gift
To see things as they are, halved by a darkness
From which they cannot shift.

Nights in the Gardens of Port of Spain

Night, our black summer, simplifies her smells
into a village; she assumes the impenetrable

musk of the Negro, grows secret as sweat,
her alleys odorous with shucked oyster shells,

coals of gold oranges, braziers of melon.
Commerce and tambourines increase her heat.

Hellfire or the whorehouse: crossing Park Street,
a surf of sailors' faces crests, is gone

with the sea's phosphorescence; the *boîtes de nuit*
twinkle like fireflies in her thick hair.

Blinded by headlamps, deaf to taxi klaxons,
she lifts her face from the cheap, pitch-oil flare

towards white stars, like cities, flashing neon,
burning to be the bitch she will become.

As daylight breaks the Indian turns his tumbril
of hacked, beheaded coconuts towards home.

Crusoe's Island

I

The chapel's cowbell
Like God's anvil
Hammers ocean to a blinding shield;
Fired, the sea grapes slowly yield
Bronze plates to the metallic heat.

Red, corrugated-iron
Roofs roar in the sun.
The wiry, ribbed air
Above earth's open kiln
Writhes like a child's vision
Of hell, but nearer, nearer.

Below, the picnic plaid
Of Scarborough is spread
To a blue, perfect sky,
Dome of our hedonist philosophy.
Bethel and Canaan's heart
Lies open like a psalm.
I labour at my art.
My father, God, is dead.

Past thirty now I know
To love the self is dread
Of being swallowed by the blue
Of heaven overhead
Or rougher blue below.
Some lesion of the brain
From art or alcohol
Flashes this fear by day:
As startling as his shadow
Grows to the castaway.

Upon this rock the bearded hermit built
His Eden:
Goats, corn crop, fort, parasol, garden,
Bible for Sabbath, all the joys
But one
Which sent him howling for a human voice.
Exiled by a flaming sun
The rotting nut, bowled in the surf,
Became his own brain rotting from the guilt
Of heaven without his kind,
Crazed by such paradisal calm
The spinal shadow of a palm
Built keel and gunwale in his mind.

The second Adam since the fall,
His germinal
Corruption held the seed
Of that congenital heresy that men fail
According to their creed.
Craftsman and castaway,
All heaven in his head,

He watched his shadow pray
Not for God's love but human love instead.

    II

We came here for the cure
Of quiet in the whelk's centre,
From the fierce, sudden quarrel,
From kitchens where the mind,
Like bread, disintegrates in water,
To let a salt sun scour
The brain as harsh as coral,
To bathe like stones in wind,
To be, like beast or natural object, pure.

That fabled, occupational
Compassion, supposedly inherited with the gift
Of poetry, had fed
With a rat's thrift on faith, shifted
Its trust to corners, hoarded
Its mania like bread,
Its brain a white, nocturnal bloom
That in a drunken, moonlit room
Saw my son's head
Swaddled in sheets
Like a lopped nut, lolling in foam.

O love, we die alone!
I am borne by the bell
Backward to boyhood
To the grey wood
Spire, harvest and marigold,

To those whom a cruel
Just God could gather
To His blue breast, His beard
A folding cloud,
As He gathered my father.
Irresolute and proud,
I can never go back.

I have lost sight of hell,
Of heaven, of human will,
My skill
Is not enough,
I am struck by this bell
To the root.
Crazed by a racking sun,
I stand at my life's noon,
On parched, delirious sand
My shadow lengthens.

III

Art is profane and pagan,
The most it has revealed
Is what a crippled Vulcan
Beat on Achilles' shield.
By these blue, changing graves
Fanned by the furnace blast
Of heaven, may the mind
Catch fire till it cleaves
Its mould of clay at last.

Now Friday's progeny,
The brood of Crusoe's slave,
Black little girls in pink
Organdy, crinolines,
Walk in their air of glory
Beside a breaking wave;
Below their feet the surf
Hisses like tambourines.

At dusk, when they return
For vespers, every dress
Touched by the sun will burn
A seraph's, an angel's,
And nothing I can learn
From art or loneliness
Can bless them as the bell's
Transfiguring tongue can bless.

Coral

This coral's shape echoes the hand
It hollowed. Its

Immediate absence is heavy. As pumice,
As your breast in my cupped palm.

Sea-cold, its nipple rasps like sand,
Its pores, like yours, shone with salt sweat.

Bodies in absence displace their weight,
And your smooth body, like none other,

Creates an exact absence like this stone
Set on a table with a whitening rack

Of souvenirs. It dares my hand
To claim what lovers' hands have never known:

The nature of the body of another.

From

THE GULF

[1970]

From *THE CASTAWAY AND OTHER POEMS* [1965]

The Flock

The grip of winter tightening, its thinned
volleys of blue-wing teal and mallard fly
from the longbows of reeds bent by the wind,
arrows of yearning for our different sky.
A season's revolution hones their sense,
whose target is our tropic light, while I
awoke this sunrise to a violence
of images migrating from the mind.
Skeletal forest, a sepulchral knight
riding in silence at a black tarn's edge,
hooves cannonading snow
in the white funeral of the year,
ant-like across the forehead of an alp
in iron contradiction crouched
against those gusts that urge the mallards south.
Vizor'd with blind defiance of his quest,
its yearly divination of the spring.
I travel through such silence, making dark
symbols with this pen's print across snow,
measuring winter's augury by words
settling the branched mind like migrating birds,
and never question when they come or go.

The style, tension of motion and the dark,
inflexible direction of the world
as it revolves upon its centuries
with change of language, climate, customs, light,
with our own prepossession day by day
year after year with images of flight,
survive our condemnation and the sun's
exultant larks.
                    The dark impartial Arctic,
whose glaciers encased the mastodon,
froze giant minds in marble attitudes,
revolves with tireless, determined grace
upon an iron axle, though the seals
howl with inhuman cries across its ice
and pages of torn birds are blown across
whitening tundras like engulfing snow,
Till its annihilation may the mind
reflect his fixity through winter, tropic,
until that equinox when the clear eye
clouds, like a mirror, without contradiction,
greet the black wings that cross it as a blessing
like the high, whirring flock that flew across
the cold sky of this page when I began
this journey by the wintry flare of dawn,
flying by instinct to their secret places,
both for their need and for my sense of season.

A Village Life

[for John Robertson]

I

Through the wide, grey loft window,
I watched that winter morning my first snow
crusting the sill, puzzle the black,
nuzzling tom. Behind my back
a rime of crud glazed my cracked coffee cup,
a snowfall of torn poems piling up,
heaped by a rhyming spade.
Starved, on the prowl,
I was a frightened cat in that grey city.
I floated, a cat's shadow, through the black wool
sweaters, leotards, and parkas of the fire-haired,
snow-shouldered Greenwich Village *mädchen*,
homesick, my desire
crawled across snow
like smoke, for its lost fire.

All that winter I haunted
your house on Hudson Street, a tiring friend,
demanding to be taken in, drunk, and fed.
I thought winter would never end.

I cannot imagine you dead.

But that stare, frozen,
a frosted pane in sunlight,
gives nothing back by letting nothing in,
your kindness or my pity.
No self-reflection lies
within those silent, ice-blue irises,
whose image is some snow-locked mountain lake
in numb Montana.

And since that winter I have learnt to gaze
on life indifferently as through a pane of glass.

II

Your image rattled on the subway glass
is my own death mask in an overcoat;
under New York, the subterranean freight
of human souls, locked in an iron cell,
station to station cowed with swaying calm,
thunders to its end, each in its private hell,
each plumped, prime bulk still swinging by its arm
upon a hook. You're two years dead. And yet
I watch that silence spreading through our souls:
that horn-rimmed midget who consoles
his own deformity with Sartre on Genet.
Terror still eats the nerves, the Word
is gibberish, the plot Absurd.
The turnstile slots, like addicts, still consume
obols and aspirin, Charon in his grilled cell
grows vague about our crime, our destination.
Not all are silent, or endure
the enormity of silence; at one station,

somewhere off 33rd and Lexington,
a fur-wrapped matron screamed above the roar
of rattling iron. Nobody took her on,
we looked away. Such scenes
rattle our trust in nerves tuned like machines.
All drives as you remember it, the pace
of walking, running the rat race,
locked in a system, ridden by its rail,
within a life where no one dares to fail.
I watch your smile breaking across my skull,
the hollows of your face below my face
sliding across it like a pane of glass.
Nothing endures. Even in his cities
man's life is grass.
Times Square. We sigh and let off steam,
who should screech with the braking wheels, scream
like our subway-Cassandra, heaven-sent
to howl for Troy, emerge
blind from the blast of daylight, whirled
apart like papers from a vent.

III

Going away, through Queens we pass
a cemetery of miniature skyscrapers. The verge
blazes its rust, its taxi-yellow leaves. It's fall.
I stare through glass,
my own reflection there, at
empty avenues, lawns, spires, quiet
stones, where the curb's rim
wheels westward, westward, where thy bones . . .

Montana, Minnesota, your real
America, lost in tall grass, serene idyll.

Goats and Monkeys

> ... *even now, an old black ram*
> *is tupping your white ewe.*
> —OTHELLO

The owl's torches gutter. Chaos clouds the globe.
Shriek, augury! His earthen bulk
buries her bosom in its slow eclipse.
His smoky hand has charred
that marble throat. Bent to her lips,
he is Africa, a vast sidling shadow
that halves your world with doubt.
"Put out the light," and God's light is put out.

That flame extinct, she contemplates her dream
of him as huge as night, as bodiless,
as starred with medals, like the moon
a fable of blind stone.
Dazzled by that bull's bulk against the sun
of Cyprus, couldn't she have known
like Pasiphaë, poor girl, she'd breed horned monsters?
That like Eurydice, her flesh a flare
travelling the hellish labyrinth of his mind
his soul would swallow hers?

Her white flesh rhymes with night. She climbs, secure.

Virgin and ape, maid and malevolent Moor,
their immortal coupling still halves our world.
He is your sacrificial beast, bellowing, goaded,
a black bull snarled in ribbons of its blood.
And yet, whatever fury girded
on that saffron-sunset turban, moon-shaped sword
was not his racial, panther-black revenge
pulsing her chamber with raw musk, its sweat,
but horror of the moon's change,
of the corruption of an absolute,
like a white fruit
pulped ripe by fondling but doubly sweet.

And so he barbarously arraigns the moon
for all she has beheld since time began,
for his own night-long lechery, ambition,
while barren innocence whimpers for pardon.
And it is still the moon, she silvers love,
limns lechery, and stares at our disgrace.
Only annihilation can resolve
the pure corruption in her dreaming face.

A bestial, comic agony. We harden
with mockery at this blackamoor
who turns his back on her, who kills
her element, night; his grief
farcically knotted in a handkerchief,
a sibyl's
prophetically stitched remembrancer
webbed and embroidered with the zodiac,
this mythical, horned beast who's no more
monstrous for being black.

Laventille

[for V. S. Naipaul]

To find the Western Path
Through the Gates of Wrath—
—BLAKE

It huddled there
steel tinkling its blue painted metal air,
tempered in violence, like Rio's favelas,

with snaking, perilous streets whose edges fell as
its Episcopal turkey-buzzards fall
from its miraculous hilltop

shrine,
down the impossible drop
to Belmont, Woodbrook, Maraval, St. Clair

that shine
like peddlers' tin trinkets in the sun.
From a harsh

shower, its gutters growled and gargled wash
past the Youth Centre, past the water catchment,
a rigid children's carousel of cement;

we climbed where lank electric
lines and tension cables linked its raw brick
hovels like a complex feud,

where the inheritors of the middle passage stewed,
five to a room, still clamped below their hatch,
breeding like felonies,

whose lives revolve round prison, graveyard, church.
Below bent breadfruit trees
in the flat, coloured city, class

escalated into structures still,
merchant, middleman, magistrate, knight. To go downhill
from here was to ascend.

The middle passage never guessed its end.
This is the height of poverty
for the desperate and black;

climbing, we could look back
with widening memory
on the hot, corrugated-iron sea
whose horrors we all

shared. The salt blood knew it well,
you, me, Samuel's daughter, Samuel,
and those ancestors clamped below its grate.

And climbing steeply past the wild
gutters, it shrilled
in the blood, for those who suffered, who were killed,

and who survive.
What other gift was there to give
as the godparents of his unnamed child?

Yet outside the brown annex of the church, the
stifling odour of bay rum and talc, the particular,
neat sweetness of the crowd distressed

that sense. The black, fawning verger,
his bow tie akimbo, grinning, the clown-gloved
fashionable wear of those I deeply loved

once, made me look on with hopelessness and rage
at their new, apish habits, their excess
and fear, the possessed, the self-possessed;

their perfume shrivelled to a childhood fear
of Sabbath graveyards, christenings, marriages,
that muggy, steaming, self-assuring air

of tropical Sabbath afternoons. And in
the church, eyes prickling with rage,
the children rescued from original sin

by their Godfather since the middle passage,
the supercilious brown curate, who intones,
healing the guilt in these rachitic bones,
twisting my love within me like a knife:
"across the troubled waters of this life . . ."

Which of us cares to walk
even if God wished
those retching waters where our souls were fished

for this new world? Afterwards, we talk
in whispers, close to death
among these stones planted on alien earth.

Afterwards,
the ceremony, the careful photograph
moved out of range before the patient tombs,

we dare a laugh,
ritual, desperate words,
born like these children from habitual wombs,

from lives fixed in the unalterable groove
of grinding poverty. I stand out on a balcony
and watch the sun pave its flat, golden path

across the roofs, the aerials, cranes, the tops
of fruit trees crawling downward to the city.
Something inside is laid wide like a wound,

some open passage that has cleft the brain,
some deep, amnesiac blow. We left
somewhere a life we never found,

customs and gods that are not born again,
some crib, some grille of light
clanged shut on us in bondage, and withheld

us from that world below us and beyond,
and in its swaddling cerements we're still bound.

Verandah

[*for Ronald Bryden*]

Grey apparitions at verandah ends
like smoke, divisible, but one
your age is ashes, its coherence gone,

Planters whose tears were marketable gum, whose voices
scratch the twilight like dried fronds
edged with reflection,

Colonels, hard as the commonwealth's greenheart,
middlemen, usurers whose art
kept an empire in the red,

Upholders of Victoria's china seas
lapping embossed around a drinking mug,
bully-boy roarers of the empire club,

To the tarantara of the bugler, the sunset furled
round the last post,
the "flamingo colours" of a fading world,

A ghost steps from you, my grandfather's ghost!
Uprooted from some rainy English shire,
you sought your Roman

End in suicide by fire.
Your mixed son gathered your charred blackened bones
in a child's coffin.

And buried them himself on a strange coast.
Sire,
why do I raise you up? Because

Your house has voices, your burnt house
shrills with unguessed, lovely inheritors,
your genealogical roof tree, fallen, survives,
like seasoned timber through green, little lives.

I ripen towards your twilight, sir, that dream
where I am singed in that sea-crossing, steam
towards that vaporous world, whose souls,

Like pressured trees, brought diamonds out of coals.
The sparks pitched from your burning house are stars.
I am the man my father loved and was.

I climb the stair
and stretch a darkening hand to greet those friends
who share with you the last inheritance
of earth, our shrine and pardoner,

grey, ghostly loungers at verandah ends.

God Rest Ye Merry, Gentlemen

Splitting from Jack Delaney's, Sheridan Square,
that winter night, stewed, seasoned in bourbon,
my body kindled by the whistling air
snowing the Village that Christ was reborn,
I lurched like any lush by his own glow
across towards Sixth, and froze before the tracks
of footprints bleeding on the virgin snow.
I tracked them where they led across the street
to the bright side, entering the wax-
sealed smell of neon, human heat,
some all-night diner with its wise-guy cook,
his stub thumb in my bowl of stew, and one
man's pulped and beaten face, its look
acknowledging all that, white-dark outside,
was possible: some beast prowling the block,
something fur-clotted, running wild
beyond the boundary of will. Outside,
more snow had fallen. My heart charred.
I longed for darkness, evil that was warm.
Walking, I'd stop and turn. What had I heard
wheezing behind my heel with whitening breath?
Nothing. Sixth Avenue yawned wet and wide.
The night was white. There was nowhere to hide.

Crusoe's Journal

> *I looked now upon the world as a thing remote, which I*
> *had nothing to do with, no expectation from, and, indeed,*
> *no desires about. In a word, I had nothing indeed*
> *to do with it, nor was ever like to have; so I thought*
> *it looked as we may perhaps look upon it hereafter,*
> *viz., as a place I had lived in but was come out*
> *of it; and well might I say, as Father Abraham*
> *to Dives, "Between me and thee is a great gulf fixed."*
> —ROBINSON CRUSOE

Once we have driven past Mundo Nuevo trace
      safely to this beach house
perched between ocean and green, churning forest
      the intellect appraises
objects surely, even the bare necessities
      of style are turned to use,
like those plain iron tools he salvages
      from shipwreck, hewing a prose
as odorous as raw wood to the adze;
      out of such timbers
came our first book, our profane Genesis
      whose Adam speaks that prose
which, blessing some sea-rock, startles itself
      with poetry's surprise,

in a green world, one without metaphors;
        like Christofer he bears
in speech mnemonic as a missionary's
        the Word to savages,
its shape an earthen, water-bearing vessel's
        whose sprinkling alters us
into good Fridays who recite His praise,
        parroting our master's
style and voice, we make his language ours,
        converted cannibals
we learn with him to eat the flesh of Christ.

All shapes, all objects multiplied from his,
        our ocean's Proteus;
in childhood, his derelict's old age
        was like a god's. (Now pass
in memory, in serene parenthesis,
        the cliff-deep leeward coast
of my own island filing past the noise
        of stuttering canvas,
some noon-struck village, Choiseul, Canaries,
        crouched crocodile canoes,
a savage settlement from Henty's novels,
        Marryat or R.L.S.,
with one boy signalling at the sea's edge,
        though what he cried is lost.)
So time, that makes us objects, multiplies
        our natural loneliness.

For the hermetic skill, that from earth's clays
        shapes something without use,

and, separate from itself, lives somewhere else,
        sharing with every beach
a longing for those gulls that cloud the cays
        with raw, mimetic cries,
never surrenders wholly, for it knows
        it needs another's praise
like hoar, half-cracked Ben Gunn, until it cries
        at last, "O happy desert!"
and learns again the self-creating peace
        of islands. So from this house
that faces nothing but the sea, his journals
        assume a household use;
we learn to shape from them, where nothing was
        the language of a race,
and since the intellect demands its mask
        that sun-cracked, bearded face
provides us with the wish to dramatize
        ourselves at nature's cost,
to attempt a beard, to squint through the sea-haze,
        posing as naturalists,
drunks, castaways, beachcombers, all of us
        yearn for those fantasies
of innocence, for our faith's arrested phase
        when the clear voice
startled itself saying "water, heaven, Christ,"
        hoarding such heresies as
God's loneliness moves in His smallest creatures.

Lampfall

Closest at lampfall
Like children, like the moth-flame metaphor,
The Coleman's humming jet at the sea's edge
A tuning fork for our still family choir
Like Joseph Wright of Derby's astrological lecture
Casts rings of benediction round the aged.
I never tire of ocean's quarrelling,
Its silence, its raw voice,
Nor of these half-lit windy leaves, gesticulating higher
"Rejoice, rejoice..."

But there's an old fish, a monster
Of primal fiction that drives barrelling
Undersea, too old to make a splash,
To which I'm hooked!
Through daydream, through nightmare trolling
Me so deep that no lights flash
There but the plankton's drifting, phosphorescent stars.

I see with its aged eyes,
Its dead green, glaucous gaze,
And I'm elsewhere, far as
I shall ever be from you whom I behold now,
Dear family, dear friends, by this still glow,
The lantern's ring that the sea's

Never extinguished.
Your voices curl in the shell of my ear.

All day you've watched
The sea-rock like a loom
Shuttling its white wool, sheer Penelope!
The coals lit, the sky glows, an oven.
Heart into heart carefully laid
Like bread.
This is the fire that draws us by our dread
Of loss, the furnace door of heaven.

At night we have heard
The forest, an ocean of leaves, drowning her children,
Still, we belong here. There's Venus. We are not yet lost.

Like you, I preferred
The firefly's starlike little
Lamp, mining, a question,
To the highway's brightly multiplying beetles.

Codicil

Schizophrenic, wrenched by two styles,
one a hack's hired prose, I earn
my exile. I trudge this sickle, moonlit beach for miles,

tan, burn
to slough off
this love of ocean that's self-love.

To change your language you must change your life.

I cannot right old wrongs.
Waves tire of horizon and return.
Gulls screech with rusty tongues

Above the beached, rotting pirogues,
they were a venomous beaked cloud at Charlotteville.

Once I thought love of country was enough,
now, even if I chose, there's no room at the trough.

I watch the best minds root like dogs
for scraps of favour.
I am nearing middle

age, burnt skin
peels from my hand like paper, onion-thin,
like Peer Gynt's riddle.

At heart there's nothing, not the dread
of death. I know too many dead.
They're all familiar, all in character,

even how they died. On fire,
the flesh no longer fears that furnace mouth
of earth,

that kiln or ashpit of the sun,
nor this clouding, unclouding sickle moon
whitening this beach again like a blank page.

All its indifference is a different rage.

From THE GULF AND OTHER POEMS
[1969]

Mass Man

Through a great lion's head clouded by mange
a black clerk growls.
Next, a gold-wired peacock withholds a man,
a fan, flaunting its oval, jewelled eyes;
What metaphors!
What coruscating, mincing fantasies!

Hector Mannix, waterworks clerk, San Juan, has entered a lion,
Boysie, two golden mangoes bobbing for breastplates, barges
like Cleopatra down her river, making style.
"Join us," they shout. "Oh God, child, you can't dance?"
But somewhere in that whirlwind's radiance
a child, rigged like a bat, collapses, sobbing.

But I am dancing, look, from an old gibbet
my bull-whipped body swings, a metronome!
Like a fruit bat dropped in the silk-cotton's shade,
my mania, my mania is a terrible calm.

Upon your penitential morning,
some skull must rub its memory with ashes,
some mind must squat down howling in your dust,
some hand must crawl and recollect your rubbish,
someone must write your poems.

Exile

Wind-haired, mufflered
against dawn, you watched the herd
of migrants ring the deck
from steerage. Only the funnel
bellowing, the gulls who peck
waste from the ploughed channel
knew that you had not come
to England; you were home.

Even her wretched weather
was poetry. Your scarred leather
suitcase held that first
indenture, to her Word,
but, among cattle docking, that rehearsed
calm meant to mark you from the herd
shook, calf-like, in her cold.

Never to go home again,
for this was home! The windows
leafed through history to the beat
of a school ballad, but the train
soon changed its poetry to the prose
of narrowing, pinched eyes you could not enter,
to the gas ring, the ringing Student's Centre,
to the soiled, icy sheet.

One night, near rheum-eyed windows
your memory kept pace with winter's
pages, piled in drifts,
till spring, which slowly lifts
the heart, broke into prose
and suns you had forgotten
blazoned from barrows.

And earth began to look
as you remembered her,
herons, like seagulls, flock-
ed to the salted furrow,
the bellowing, smoky bullock
churned its cane sea,
a world began to pass
through your pen's eye,
between bent grasses and one word
for the bent rice.

And now, some phrase
caught in the parenthesis
of highway quietly states
its title, and an ochre trace
of flags and carat huts opens
at Chapter One,
the bullock's strenuous ease is mirrored
in a clear page of prose,
a forest is compressed in a blue coal,
or burns in graphite fire,
invisibly your ink nourishes
leaf after leaf the furrowed villages
where the smoke flutes

and the brittle pages
of the Ramayana stoke the mulch fires,

the arrowing, metal
highways head nowhere,
the tabla and the sitar amplified,
the Path unrolling like a dirty bandage,
the cinema-hoardings leer
in language half the country cannot read.

Yet, when dry winds rattle
the flags whose bamboo lances bend
to Hanuman, when, like chattel
folded in a cloth knot, the debased
brasses are tremblingly placed
on flaking temple lintels,
when the god stamps his bells
and smoke writhes its blue arms
for your lost India,

the old men, threshing rice,
rheum-eyed, pause,
their brown gaze flecked with chaff,
their loss chafed by the raw
whine of the cinema-van calling the countryside
to its own dark devotions,
summoning the drowned from oceans
of deep cane. The hymn
to Mother India whores its lie.
Your memory walks by its soft-spoken
path, as flickering, broken,
Saturday jerks past like a cheap film.

Homage to Edward Thomas

Formal, informal, by a country's cast
topography delineates its verse,
erects the classic bulk, for rigid contrast
of sonnet, rectory or this manor house
dourly timbered against these sinuous
Downs, defines the formal and informal prose
of Edward Thomas's poems, which make this garden
return its subtle scent of Edward Thomas
in everything here hedged or loosely grown.
Lines which you once dismissed as tenuous
because they would not howl or overwhelm,
as crookedly grave-bent, or cuckoo-dreaming,
seeming dissoluble as this Sussex down
harden in their indifference, like this elm.

The Gulf

[*for Jack and Barbara Harrison*]

I

The airport coffee tastes less of America.
Sour, unshaven, dreading the exertion
of tightening, racked nerves fuelled with liquor,

some smoky, resinous bourbon,
the body, buckling at its casket hole,
a roar like last night's blast racing its engines,

watches the fumes of the exhausted soul
as the trans-Texas jet, screeching, begins
its flight and friends diminish. So, to be aware

of the divine union the soul detaches
itself from created things. "We're in the air,"
the Texan near me grins. All things: these matches

from LBJ's campaign hotel, this rose
given me at dawn in Austin by a child,
this book of fables by Borges, its prose

a stalking, moonlit tiger. What was willed
on innocent, sun-streaked Dallas, the beast's claw
curled round that hairspring rifle is revealed

on every page as lunacy or feral law;
circling that wound we leave Love Field.
Fondled, these objects conjure hotels,

quarrels, new friendships, brown limbs
nakedly moulded as these autumn hills
memory penetrates as the jet climbs

the new clouds over Texas; their home means
an island suburb, forest, mountain water;
they are the simple properties for scenes

whose joy exhausts like grief, scenes where we learn,
exchanging the least gifts, this rose, this napkin,
that those we love are objects we return,

that this lens on the desert's wrinkled skin
has priced our flesh, all that we love in pawn
to that brass ball, that the gifts, multiplying,

clutter and choke the heart, and that I shall
watch love reclaim its things as I lie dying.
My very flesh and blood! Each seems a petal

shrivelling from its core. I watch them burn,
by the nerves' flare I catch their skeletal
candour! Best never to be born,

the great dead cry. Their works shine on our shelves,
by twilight tour their gilded gravestone spines,
and read until the lamplit page revolves

to a white stasis whose detachment shines
like a propeller's rainbowed radiance.
Circling like us; no comfort for their loves!

II

The cold glass darkens. Elizabeth wrote once
that we make glass the image of our pain;
I watch clouds boil past the cold, sweating pane

above the Gulf. All styles yearn to be plain
as life. The face of the loved object under glass
is plainer still. Yet, somehow, at this height,

above this cauldron boiling with its wars,
our old earth, breaking to familiar light,
that cloud-bound mummy with self-healing scars

peeled of her cerements again looks new;
some cratered valley heals itself with sage,
through that grey, fading massacre a blue

lighthearted creek flutes of some siege
to the amnesia of drumming water.
Their cause is crystalline: the divine union

of these detached, divided states, whose slaughter
darkens each summer now, as one by one,
the smoke of bursting ghettos clouds the glass

down every coast where filling station signs
proclaim the Gulf, an air, heavy with gas,
sickens the state, from Newark to New Orleans.

III

Yet the South felt like home. Wrought balconies,
the sluggish river with its tidal drawl,
the tropic air charged with the extremities

of patience, a heat heavy with oil,
canebrakes, that legendary jazz. But fear
thickened my voice, that strange, familiar soil

prickled and barbed the texture of my hair,
my status as a secondary soul.
The Gulf, your gulf, is daily widening,

each blood-red rose warns of that coming night
when there's no rock cleft to go hidin' in
and all the rocks catch fire, when that black might,

their stalking, moonless panthers turn from Him
whose voice they can no more believe, when the black X's
mark their passover with slain seraphim.

IV

The Gulf shines, dull as lead. The coast of Texas
glints like a metal rim. I have no home
as long as summer bubbling to its head

boils for that day when in the Lord God's name
the coals of fire are heaped upon the head
of all whose gospel is the whip and flame,

age after age, the uninstructing dead.

Elegy

Our hammock swung between Americas,
we miss you, Liberty. Che's
bullet-riddled body falls,
and those who cried, the Republic must first die
to be reborn, are dead,
the freeborn citizen's ballot in the head.
Still, everybody wants to go to bed
with Miss America. And, if there's no bread,
let them eat cherry pie.

But the old choice of running, howling, wounded
wolf-deep in her woods,
while the white papers snow on
genocide is gone;
no face can hide
its public, private pain,
wincing, already statued.

Some splintered arrowhead lodged in her brain
sets the black singer howling in his bear trap,
shines young eyes with the brightness of the mad,
tires the old with her residual sadness;
and yearly lilacs in her dooryards bloom,
and the cherry orchard's surf
blinds Washington and whispers

to the assassin in his furnished room
of an ideal America, whose flickering screens
show, in slow herds, the ghosts of the Cheyennes
scuffling across the staked and wired plains
with whispering, rag-bound feet,

while the farm couple framed in their Gothic door
like Calvin's saints, waspish, pragmatic, poor,
gripping the devil's pitchfork
stare rigidly towards the immortal wheat.

6 June 1968

Blues

Those five or six young guys
hunched on the stoop
that oven-hot summer night
whistled me over. Nice
and friendly. So, I stop.
MacDougal or Christopher
Street in chains of light.

A summer festival. Or some
saint's. I wasn't too far from
home, but not too bright
for a nigger, and not too dark.
I figured we were all
one, wop, nigger, jew,
besides, this wasn't Central Park.
I'm coming on too strong? You figure
right! They beat this yellow nigger
black and blue.

Yeah. During all this, scared
in case one used a knife,
I hung my olive-green, just-bought
sports coat on a fire plug.
I did nothing. They fought
each other, really. Life

gives them a few kicks,
that's all. The spades, the spicks.

My face smashed in, my bloody mug
pouring, my olive-branch jacket saved
from cuts and tears,
I crawled four flights upstairs.
Sprawled in the gutter, I
remember a few watchers waved
loudly, and one kid's mother shouting
like "Jackie" or "Terry,"
"now that's enough!"
It's nothing really.
They don't get enough love.

You know they wouldn't kill
you. Just playing rough,
like young America will.
Still, it taught me something
about love. If it's so tough,
forget it.

Air

> *There has been romance, but it has been the romance of pirates and outlaws. The natural graces of life do not show themselves under such conditions. There are no people there in the true sense of the word, with a character and purpose of their own.*
> —FROUDE, *The Bow of Ulysses*

The unheard, omnivorous
jaws of this rain forest
not merely devour all
but allow nothing vain;
they never rest,
grinding their disavowal
of human pain.

Long, long before us,
those hot jaws, like an oven
steaming, were open
to genocide; they devoured
two minor yellow races, and
half of a black;
in the Word made flesh of God
all entered that gross un-
discriminating stomach;

the forest is unconverted,
because that shell-like noise
which roars like silence, or
ocean's surpliced choirs
entering its nave, to a censer
of swung mist, is not
the rustling of prayer
but nothing; milling air,
a faith, infested, cannibal,
which eats gods, which devoured
the god-refusing Carib, petal
by golden petal, then forgot,
and the Arawak
who leaves not the lightest fern-trace
of his fossil to be cultured
by black rock,

but only the rusting cries
of a rainbird, like a hoarse
warrior summoning his race
from vaporous air
between this mountain ridge
and the vague sea
where the lost exodus
of corials sunk without trace—

there is too much nothing here.

Guyana

I

The surveyor straightens from his theodolite.
"Spirit-level," he scrawls, and instantly
the ciphers staggering down their columns
are soldier ants, their panic radiating in the shadow
of a new god arriving over Aztec anthills.

The sun has sucked his brain pith-dry.
His vision whirls with dervishes, he is dust.
Like an archaic photographer, hooded in shade,
he crouches, screwing a continent to his eye.

The vault that balances on a grass blade,
the nerve-cracked ground too close for the word "measureless,"
for the lost concept, "man,"
revolve too slowly for the fob watch of his world,
for the tidal markings of the five-year plan.

Ant-sized to God, god to an ant's eyes,
shouldering science he begins to tread
himself, a world that must be measured in three days.

The frothing shallows of the river,
the forest so distant that it tires of blue,
the merciless idiocy of green, green . . .

a shape dilates towards him through the haze.

II/ *The Bush*

Together they walked through a thickness pinned with birds
silent as rags, grackles and flycatchers mostly,
shaking words from their heads,

their beaks aimed at one target, the clotting sun.
Tight, with the tension of arrows.

Dark climbed their knees until their heads were dark.
The wind, wave-muscled, kept its steady mowing.
Thoughts fell from him like leaves.

He followed, that was all,
his mind, one step behind,
pacing the poem, going where it was going.

III/ *The White Town*

"Man, all the men in that damned country mad!"
There was the joke on W. and Mayakovsky.
There was the charred bush of a man found in the morning,
there was the burgher's glare of whitewashed houses
outstaring guilt,
                there was the anthropologist
dropping on soft pads from the thorn branches
to the first stance hearing the vowels
fur in his throat the hoarse
pebbles of consonants rattling his parched gullet,
there was the poet howling in vines of syntax

and the surveyor
dumbstruck by a stone;
                            at noon, the ferment
of white air, lilies and canal water
heavy as bush rum, then amber
saddening twilights without ice.
A fist should smash the glare of skylight open.
In the asylum the prisoners slept like snakes,
their eyes wide open.
                    They wait.

All of us wait.

   IV/*The Falls*

Their barrelling roar would open like a white oven
for him,
who was a spirit now, who could not burn or drown.

Surely in that "smoke that thundered" there was a door—
but the noise boiled to the traffic of a white town
of bicycles, pigeons, bells, smoke, trains at the rush hour

revolving to this roar.
He was a flower,
weightless. He would float down.

   V/*A Map of the Continent*

The lexicographer in his cell records the life and death of books;
the naked buck waits at the edge of the world.

One hefts a pen, the other a bone spear;
between them curls a map,
between them curl the vigorous, rotting leaves,

shelves forested with titles, trunks that wait for names—
it pierces knowledge, the spear-flash!
the fish thrashing green air
on a pen's hook,

above the falls reciting its single flower.

The lexicographer's lizard eyes are curled
in sleep. The Amazonian Indian enters them.

Between the Rupununi and Borges,
between the fallen pen tip and the spearhead
thunders, thickens, and shimmers the one age of the world.

    VI/ *A Georgetown Journal*

    1

Begun, with its own impulse of destruction,
this elegy that chokes its canals
like the idle, rotting lilies of this frontier,
its lines that rust, however shiningly they thrust forward,
    like the elementary railway
besieging the whitewashed city
that reminds the poet on his balcony of thunder.

Begun, with a brown heron,
like the one I named for an actor,
its emblem answering a question with a question:

"What bird is that,
whose is that woman,
what will become of their country?"

If the neck of the heron is condemned to its question,
if the woman is silent,
and if, at the most appropriate hour
of a rose-scrim twilight budding with onion domes
like the gaze of clerkish guerrillas hazed by an epoch,
if nothing comes,
if no one ever escapes,
if the shoreline longs sadly for spires,
there is nothing left for us
but to make these coarse lilies lotuses,
for filth to contemplate its own reflection.

Cycle bells startle the pigeons.
The air has been cleared of hawks,
and the bourgeois gurgling like canals
reminisce over carrion.

Spires walk the sea-wall.

The wind unwraps them to wires.
They recede, skeletal, skeletal,
the streets have grown ordinary as heroes.

And the prose of polemics grows, spreading lianas of syntax
for the rootless surveyor,
the thunderous falls have been measured,
the thickening girth of the continent has been buttoned
till a man knows his weight to the stone,

his worth to the inch,
yet imagines he hears in his hair
the rain horses crossing savannahs
and his pores prickle like water.

The towns are clogged at their edges,
a glutinous dialect chokes the slum's canals,
and the white, finical houses
lift their lace skirts, stepping over the creeks.

Hawsers have lifted the country on delicate ankles.
The dead face of an orator revolves by lamplight,
the glazed scar itches for blood.

The girl waits in the wings, heron-still;
she will rise to the roar of the playhouse
its applauding cataract,

and the train rusts, travelling to a few sad sparks,
and the muck, and the tins, and the sogged placards choke
the sad, motionless green of the canals.

   2
So, safest, I had unimagined time;
thus we forget our element
like a fish that gasps with surprise on the nib of a hook.

There was always death,
but that came in the cheapest pricklings,
in old songs, in the amazed fading of letters,
in the change in one's penmanship:
how an *l* wavers like a single lily shaken

by a stone, how an *r* reaches rightly, to touch
some vertical end,

and at startling moments, the rattle of a kite
on a pluperfect sky.
Now words like "azure," for instance, suddenly touch,
such homilies as "infinite" momentarily burn,
and for these lined eyes to widen
the heart, when to have written "heart"
was to know a particular spasm—
how an old rock could spout such crystalline gibberish
amuses me, like the exact dancing of machines.

Nothing could turn my head, not the night moth,
like a nun beating her prison, but now pain comes
where I least expect it:
in the hissing of bicycle tires on drizzled asphalt,
in the ambush of little infinities
as supple with longing as the word "horizon."
Sad is the felon's love for the scratched wall,
beautiful the exhaustion of old towels,
and the patience of dented saucepans
seems mortally comic.

All these predictions do not disappoint but bring us nearer.
They uphold history like a glass of water.
If the poem begins to shrivel
I no longer distend my heart,
for I know how profound is the folding of a napkin
by a woman whose hair will go white,
age, that says more than an ocean,
I know how final is the straightening of a sheet

between lovers who have never lain, the heartbreaking curve
of a woman, her back bent, concerned
with the finical precisions of farewell.

    3

And there I entered your green, sibilant Russias,
those canes that like wheat must blacken after harvest,
and I honoured your dead, those few
arranged in postures for your great elegies,
who are what they were, not heroes, merely men.

The age will know its own name when it comes,
as love will find its breath softly expelling
"was I like this?"
with the same care, the precise exhilaration
with which the heron's foot pronounces "earth."

What if, impulsive, delicate bird,
one instinct made you rise
out of this life, into another's,
then from another's, circling to your own?
You are folded in my eyes,
whose irises will open
to a white sky with bird and woman gone.

Che

In this dark-grained news photograph, whose glare
is rigidly composed as Caravaggio's,
the corpse glows candle-white on its cold altar—

its stone Bolivian Indian butcher's slab-
stare till its waxen flesh begins to harden
to marble, to veined, white Andean iron;
from your own fear, *cabrón,* its pallor grows;

it stumbled from your doubt, and for your pardon
burnt in brown trash, far from the embalming snows.

Negatives

A news clip; the invasion of Biafra:
black corpses wrapped in sunlight
sprawled on the white glare entering what's-its-name—
the central city?
                      Someone who's white
illuminates the news behind the news,
his eyes flash with, perhaps, pity:
"The Ibos, you see, are like the Jews,
very much the situation in Hitler's Germany,
I mean the Hausas' resentment." I try to see.

I never knew you, Christopher Okigbo,
I saw you when an actor screamed "The tribes!
The tribes!" I catch
the guttering, flare-lit
faces of Ibos,
stuttering, bug-eyed
prisoners of some drumhead tribunal.

The soldiers' helmeted shadows
could have been white, and yours
one of those sun-wrapped bodies on the white road
entering ... the tribes, the tribes, their shame—
that central city, Christ, what is its name?

Landfall, Grenada

[*for Robert Head, mariner*]

Where you are rigidly anchored,
the groundswell of blue foothills, the blown canes
surging to cumuli cannot be heard;
like the slow, seamless ocean,
one motion folds the grass where you were lowered,
and the tiered sea
whose grandeurs you detested
climbs out of sound.

Its moods held no mythology
for you, it was a working place
of tonnage and ruled stars;
you chose your landfall with a mariner's
casual certainty,
calm as that race
into whose heart you harboured;
your death was a log's entry,
your suffering held the strenuous
reticence of those
whose rites are never public,
hating to impose, to offend.
Deep friend, teach me to learn
such ease, such landfall going,

such mocking tolerance of those
neat gravestone elegies
that rhyme our end.

Homecoming: Anse La Raye

[*for Garth St. Omer*]

Whatever else we learned
at school, like solemn Afro-Greeks eager for grades,
of Helen and the shades
of borrowed ancestors,
there are no rites
for those who have returned,
only, when her looms fade,
drilled in our skulls, the doom-
surge-haunted nights,
only this well-known passage

under the coconuts' salt-rusted
swords, these rotted
leathery sea-grape leaves,
the seacrabs' brittle helmets, and
this barbecue of branches, like the ribs
of sacrificial oxen on scorched sand;
only this fish-gut-reeking beach
whose frigates tack like buzzards overhead,
whose spindly, sugar-headed children race
pelting up from the shallows
because your clothes,
your posture

seem a tourist's.
They swarm like flies
round your heart's sore.

Suffer them to come,
entering your needle's eye,
knowing whether they live or die,
what others make of life will pass them by
like that far silvery freighter
threading the horizon like a toy;
for once, like them,
you wanted no career
but this sheer light, this clear,
infinite, boring, paradisal sea,
but hoped it would mean something to declare
today, I am your poet, yours,
all this you knew,
but never guessed you'd come
to know there are homecomings without home.

You give them nothing.
Their curses melt in air.
The black cliffs scowl,
the ocean sucks its teeth,
like that dugout canoe
a drifting petal fallen in a cup,
with nothing but its image,
you sway, reflecting nothing.
The freighter's silvery ghost
is gone, the children gone.
Dazed by the sun
you trudge back to the village

past the white, salty esplanade
under whose palms dead
fishermen move their draughts in shade,
crossing, eating their islands,
and one, with a politician's
ignorant, sweet smile, nods,
as if all fate
swayed in his lifted hand.

Star

If, in the light of things, you fade
real, yet wanly withdrawn
to our determined and appropriate
distance, like the moon left on
all night among the leaves, may
you invisibly delight this house;
O star, doubly compassionate, who came
too soon for twilight, too late
for dawn, may your pale flame
direct the worst in us
through chaos
with the passion of
plain day.

Cold Spring Harbor

From feather-stuffed bolsters of cloud
falling on casual linen
the small shrieks soundlessly float.
The woods are lint-wreathed. Dawn
crackles like foil to the rake
of a field mouse nibbling, nibbling
its icing. The world is unwrapped
in cotton and you would tread wool
if you opened, quietly, whitely,
this door, like an old Christmas card
turned by a child's dark hand, did
he know it was dark then,
the magical brittle branches, the white house
collared in fur, the white world of men,
its bleeding gules and its berry drops?

Two prancing, immobile white ponies
no bigger than mice pulled a carriage
across soundless hillocks of cotton;
bells hasped to their necks didn't tinkle
though you begged God to touch them to life,
some white-haired old God who'd forgotten
or no longer trusted his miracles.
What urges you now towards this white,
snow-whipped woods is not memory

of that dark child's toys, not the card
of a season, forever foreign, that went
over its ridges like a silent
sleigh. That was a child's sorrow, this is
child's play, through which you cannot go,
dumbstruck at an open door,
stunned, fearing the strange violation
(because you are missing your children)
of perfect snow.

Love in the Valley

The sun goes slowly blind.
It is this mountain, shrouding
the valley of the shadow,

widening like amnesia
evening dims the mind.
I shake my head in darkness,

it is a tree branched with cries,
a trash can full of print.
Now, through the reddening squint

of leaves leaden as eyes,
a skein of drifting hair
like a twig fallen on snow

branches the blank pages.
I bring it close, and stare
in slow vertiginous darkness,

and now I drift elsewhere,
through hostile images
of white and black, and look,

like a thaw-sniffing stallion, the head
of Pasternak emerges with its forelock,
his sinewy wrist a fetlock

pawing the frozen spring,
till his own hand has frozen
on the white page, heavy.

I ride through a white childhood
whose pines glittered with bracelets,
when I heard wolves, feared the black wood,

every wrist-aching brook
and the ice maiden
in Hawthorne's fairy book.

The hair melts into dark,
a question mark that led
where the untethered mind

strayed from its first track;
now Hardy's sombre head
upon which hailstorms broke

looms, like a weeping rock,
like wind, the tresses drift
and their familiar trace

tingles across the face
with light lashes.
I knew the depth of whiteness,

I feared the numbing kiss
of those women of winter,
Bathsheba, Lara, Tess,

whose tragedy made less
of life, whose love was more
than love of literature.

Nearing Forty

[*for John Figueroa*]

The irregular combination of fanciful invention may delight awhile by that novelty of which the common satiety of life sends us all in quest. But the pleasures of sudden wonder are soon exhausted and the mind can only repose on the stability of truth . . .
—SAMUEL JOHNSON

Insomniac since four, hearing this narrow,
rigidly metred, early-rising rain
recounting, as its coolness numbs the marrow,
that I am nearing forty, nearer the weak
vision thickening to a frosted pane,
nearer the day when I may judge my work
by the bleak modesty of middle age
as a false dawn, fireless and average,
which would be just, because your life bled for
the household truth, the style past metaphor
that finds its parallel however wretched
in simple, shining lines, in pages stretched
plain as a bleaching bedsheet under a guttering rainspout, glad for the sputter
of occasional insight; you who foresaw
ambition as a searing meteor

will fumble a damp match and, smiling, settle
for the dry wheezing of a dented kettle,
for vision narrower than a louvre's gap,
then, watching your leaves thin, recall how deep
prodigious cynicism plants its seed,
gauges our seasons by this year's end rain
which, as greenhorns at school, we'd
call conventional for convectional;
or you will rise and set your lines to work
with sadder joy but steadier elation,
until the night when you can really sleep,
measuring how imagination
ebbs, conventional as any water clerk
who weighs the force of lightly falling rain,
which, as the new moon moves it, does its work
even when it seems to weep.

The Walk

After hard rain the eaves repeat their beads,
those trees exhale your doubt like mantled tapers,
drop after drop, like a child's abacus
beads of cold sweat file from high-tension wires,

pray for us, pray for this house, borrow your neighbour's
faith, pray for this brain that tires,
and loses faith in the great books it reads;
after a day spent prone, hemorrhaging poems,

each phrase peeled from the flesh in bandages,
arise, stroll on under a sky
sodden as kitchen laundry,

while the cats yawn behind their window frames,
lions in cages of their choice,
no further, though, than your last neighbour's gates
figured with pearl. How terrible is your own

fidelity, O heart, O rose of iron!
When was your work more like a housemaid's novel,
some drenched soap opera which gets
closer than yours to life? Only the pain,

the pain is real. Here's your life's end,
a clump of bamboos whose clenched
fist loosens its flowers, a track
that hisses through the rain-drenched

grove: abandon all, the work,
the pain of a short life. Startled, you move;
your house, a lion rising, paws you back.

ANOTHER LIFE

[1973]

One
THE DIVIDED CHILD

> *An old story goes that Cimabue was struck with admiration when he saw the shepherd boy, Giotto, sketching sheep. But, according to the true biographies, it is never the sheep that inspire a Giotto with the love of painting: but, rather, his first sight of the paintings of such a man as Cimabue. What makes the artist is the circumstance that in his youth he was more deeply moved by the sight of works of art than by that of the things which they portray.*
>
> MALRAUX
> *Psychology of Art*

Chapter 1

I

Verandahs, where the pages of the sea
are a book left open by an absent master
in the middle of another life—
I begin here again,
begin until this ocean's
a shut book, and like a bulb
the white moon's filaments wane.

Begin with twilight, when a glare
which held a cry of bugles lowered
the coconut lances of the inlet,
as a sun, tired of empire, declined.
It mesmerized like fire without wind,
and as its amber climbed
the beer-stein ovals of the British fort
above the promontory, the sky
grew drunk with light.
                      There
was your heaven! The clear
glaze of another life,
a landscape locked in amber, the rare
gleam. The dream
of reason had produced its monster:
a prodigy of the wrong age and colour.

All afternoon the student
with the dry fever of some draughtsman's clerk
had magnified the harbour, now twilight
eager to complete itself,
drew a girl's figure to the open door
of a stone boathouse with a single stroke, then fell
to a reflecting silence. This silence waited
for the verification of detail:
the gables of the St. Antoine Hotel
aspiring from jungle, the flag
at Government House melting its pole,
and for the tidal amber glare to glaze
the last shacks of the Morne till they became
transfigured sheerly by the student's will,
a cinquecento fragment in gilt frame.

The vision died,
the black hills simplified
to hunks of coal,
but if the light was dying through the stone
of that converted boathouse on the pier,
a girl, blowing its embers in her kitchen,
could feel its epoch entering her hair.

Darkness, soft as amnesia, furred the slope.
He rose and climbed towards the studio.
The last hill burned,
the sea crinkled like foil,
a moon ballooned up from the Wireless Station. O
mirror, where a generation yearned
for whiteness, for candour, unreturned.

The moon maintained her station,
her fingers stroked a chiton-fluted sea,
her disc whitewashed the shells
of gutted offices barnacling the wharves
of the burnt town, her lamp
baring the ovals of toothless façades,
along the Roman arches, as he passed
her alternating ivories lay untuned,
her age was dead, her sheet
shrouded the antique furniture, the mantel
with its plaster-of-Paris Venus, which
his yearning had made marble, half-cracked
unsilvering mirror of black servants,
like the painter's kerchiefed, ear-ringed portrait: Albertina.

Within the door, a bulb
haloed the tonsure of a reader crouched
in its pale tissue like an embryo,
the leisured gaze
turned towards him, the short arms
yawned briefly, welcome. Let us see.
Brown, balding, a lacertilian
jut to its underlip,
with spectacles thick as a glass paperweight
over eyes the hue of sea-smoothed bottle glass,
the man wafted the drawing to his face
as if dusk were myopic, not his gaze.
Then, with slow strokes, the master changed the sketch.

II

In its dimension the drawing could not trace
the sociological contours of the promontory;
once, it had been an avenue of palms
strict as Hobbema's aisle of lowland poplars,
now, levelled, bulldozed, and metalled for an airstrip,
its terraces like tree rings told its age.
There, patriarchal banyans,
bearded with vines from which black schoolboys gibboned,
brooded on a lagoon seasoned with dead leaves,
mangroves knee-deep in water
crouched like whelk pickers on brown, spindly legs
scattering red soldier crabs
scrabbling for redcoats' meat.
The groves were sawn
symmetry and contour crumbled,
down the arched barrack balconies
where colonels in the whisky-coloured light
had watched the green flash, like a lizard's tongue,
catch the last sail, tonight
row after row of orange stamps repeated
the villas of promoted civil servants.

The moon came to the window and stayed there.
He was her subject, changing when she changed,
from childhood he'd considered palms
ignobler than imagined elms,
the breadfruit's splayed
leaf coarser than the oak's,
he had prayed

nightly for his flesh to change,
his dun flesh peeled white by her lightning strokes!
Above the cemetery where
the airstrip's tarmac ended
her slow disc magnified
the life beneath her like a reading glass.

Below the bulb
a green book, laid
face downward. Moon,
and sea. He read
the spine. FIRST POEMS:
CAMPBELL. The painter
almost absently
reversed it, and began to read:

> "Holy be
> the white head of a Negro,
> sacred be
> the black flax of a black child . . ."

And from a new book,
bound in sea-green linen, whose lines
matched the exhilaration which their reader,
rowing the air around him now, conveyed,
another life it seemed would start again,
while past the droning, tonsured head
the white face
of a dead child stared from its window-frame.

III

They sang, against the rasp and cough of shovels,
against the fists of mud pounding the coffin,
the diggers' wrists rounding off every phrase,
their iron hymn, "The Pilgrims of the Night."
In the sea-dusk, the live child waited
for the other to escape, a flute
of frail, seraphic mist,
but their black, Bible-paper voices fluttered shut, silence
re-entered every mould, it wrapped the edges
of sea-eaten stone, mantled the blind
eternally gesturing angels, strengthened the flowers
with a different patience, and left
or lost its hoarse voice in the shells
that trumpeted from the graves. The world
stopped swaying and settled in its place.
A black lace glove swallowed his hand.
The engine of the sea began again.

A night-black hearse, tasselled and heavy, lugged
an evening of blue smoke across the field,
like an old wreath the mourners broke apart
and drooped like flowers over the streaked stones
deciphering dates. The gravekeeper with his lantern-jaw
(years later every lantern-swinging porter
guarding infinite rails repeated this) opened
the yellow doorway to his lodge. Wayfarer's station.
The child's journey was signed.
The ledger drank its entry.
Outside the cemetery gates life stretched from sleep.

Gone to her harvest of flax-headed angels,
of seraphs blowing pink-palated conchs,
gone, so they sang, into another light:
But was it her?
Or Thomas Alva Lawrence's dead child,
another Pinkie, in her rose gown floating?
Both held the same dark eyes,
slow, haunting coals, the same curved
ivory hand touching the breast,
as if, answering death, each whispered "Me?"

   IV

Well, everything whitens,
all that town's characters, its cast of thousands
arrested in one still!
As if a sudden flashbulb showed their deaths.
The trees, the road he walks home, a white film,
tonight in the park the children leap into statues,
their outcries round as moonlight,
their flesh like flaking stone,
poor negatives!
They have soaked too long in the basin of the mind,
they have drunk the moon-milk
that X-rays their bodies,
the bone tree shows
through the starved skins,
and one has left, too soon,
a reader out of breath,
and once that begins, how shall I tell them,
while the tired filaments of another moon,
one that was younger,
fade, with the elate extinction of a bulb?

Chapter 2

I

At every first communion, the moon
would lend her lace to a barefooted town
christened, married, and buried in borrowed white,
in fretwork borders of carpenter's Gothic,
in mansard bonnets, pleated jalousies,
when, with her laces laid aside,
she was a servant, her sign
a dry park of disconsolate palms, like brooms,
planted by the seventh Edward, Prince of Wales,
with drooping ostrich crests, ICH DIEN, I SERVE.

I sweep. I iron. The smell of drizzled asphalt
like a flatiron burning,
odours of smoke, the funereal berried ferns
that made an undertaker's parlour of our gallery.
Across the pebbled back yard, woodsmoke thins,
epiphany of ascension. The soul, like fire,
abhors what it consumes. In the upstairs rooms
smell of blue soap that puckered the black nurse's palms,
those hands which held our faces like a vase;
the coffee grinder, grumbling,
ground its teeth,
waking at six.
                The cracked egg hisses.

The sheets of Monday
are fluttering from the yard.
The week sets sail.

   II

               Maman,
only on Sundays was the Singer silent,
then
tobacco smelt stronger, was more masculine.
Sundays
the parlour smelt of uncles,
the lamp poles rang,
the drizzle shivered its maracas,
like mandolins the tightening wires of rain,
then
from striped picnic buses, *jour marron*,
gold bangles tinkled like good-morning in Guinea
and a whore's laughter opened like sliced fruit.

Maman,
you sat folded in silence,
as if your husband might walk up the street,
while in the forests the cicadas pedalled their machines,
and silence, a black maid in white,
barefooted, polished and repolished
the glass across his fading watercolours,
the dumb Victrola cabinet,
the panels and the gleam of blue-winged teal
beating the mirror's lake.
In silence,
the revered, silent objects ring like glass,

at my eyes' touch, everything tightened, tuned,
Sunday,
the dead Victrola;

Sunday, a child
breathing with lungs of bread;
Sunday, the sacred silence of machines.

Maman,
your son's ghost circles your lost house, looking in
incomprehensibly at its dumb tenants
like fishes busily inaudible behind glass,
while the carpenter's Gothic joke, A, W, A, W,
Warwick and Alix involved in its eaves
breaks with betrayal.
You stitched us clothes from the nearest elements,
made shirts of rain and freshly ironed clouds,
then, singing your iron hymn, you riveted
your feet on Monday to the old machine.

Then Monday plunged her arms up to the elbows
in a foam tub, under a blue-soap sky,
the wet fleets sailed the yard, and every bubble,
with its bent, mullioned window, opened
its mote of envy in the child's green eye
of that sovereign-headed, pink-cheeked bastard Bubbles
in the frontispiece of *Pears Cyclopedia*.
Rising in crystal spheres, world after world.

They melt from you, your sons.
Your arms grow full of rain.

III

Old house, old woman, old room,
old planes, old buckling membranes of the womb,
translucent walls,
breathe through your timbers; gasp
arthritic, curling beams,
cough in old air
shining with motes, stair
polished and repolished by the hands of strangers,
die with defiance flecking your grey eyes,
motes of a sunlit air,
your timbers humming with constellations of carcinoma,
your bed frames glowing with radium,
cold iron dilating the fever of your body,
while the galvanized iron snaps in spasms of pain,
but a house gives no outcry,
it bears the depth of forest, of ocean and mother.
Each consuming the other
with memory and unuse.

Why should we weep for dumb things?

This radiance of sharing extends to the simplest objects,
to a favourite hammer, a paintbrush, a toothless,
gum-sunken old shoe,
to the brain of a childhood room, retarded,
lobotomized of its furniture,
stuttering its inventory of accidents:
why this chair cracked,
when did the tightened scream

of that bedspring finally snap,
when did that unsilvering mirror finally
surrender her vanity,
and, in turn, these objects assess us,
that yellow paper flower with the eyes of a cat,
that stain, familiar as warts or some birthmark,
as the badge of some loved defect,

while the thorns of the bougainvillea
moult like old fingernails,
and the flowers keep falling,
and the flowers keep opening,
the allamandas' fallen bugles, but nobody charges.

Skin wrinkles like paint,
the forearm of a balustrade freckles,
crows' feet radiate
from the shut eyes of windows,
and the door, mouth clamped, reveals nothing,
for there is no secret,
there is no other secret
but a pain so alive that
to touch every ledge of that house edges a scream
from the burning wires, the nerves
with their constellation of cancer,
the beams with their star-seed of lice,
pain shrinking every room,
pain shining in every womb,
while the blind, dumb
termites, with jaws of the crabcells consume,
in silent thunder,

to the last of all Sundays,
consume.

Finger each object, lift it
from its place, and it screams again
to be put down
in its ring of dust, like the marriage finger
frantic without its ring;
I can no more move you from your true alignment,
Mother, than we can move objects in paintings.

Your house sang softly of balance,
of the rightness of placed things.

Chapter 3

I

Each dusk the leaf flared on its iron tree,
the lamplighter shouldered his ladder, a sickle
of pale light fell on the curb.

The child tented his cotton nightdress tight
across his knees. A kite
whose twigs showed through. Twilight
enshrined the lantern of his head.

Hands swing him heavenward.
The candle's yellow leaf next to his bed
re-letters *Tanglewood Tales* and Kingsley's *Heroes*,
gilding their backs,

the ceiling reels with magic lantern shows.
The black lamplighter with Demeter's torch
ignites the iron trees above the shacks.
Boy! Who was Ajax?

II

Ajax,
    lion-coloured stallion from Sealey's stable,
    by day a cart horse, a thoroughbred

on race days, once a year,
   plunges the thunder of his neck, and sniffs
   above the garbage smells, the scent
   of battle, and the shouting,
   he saith among the kitchen peels, "Aha!"
   debased, bored animal,
   its dung cakes pluming, gathers
   the thunder of its flanks, and drags
   its chariot to the next block, where

Berthilia,
   the frog-like, crippled crone,
   a hump on her son's back, is carried
   to her straw mat, her day-long perch,
   Cassandra, with her drone unheeded.
   Her son, Pierre, carries night-soil in buckets,
   she spurs him like a rider,
   horsey-back, horsey-back;
   when he describes his cross he sounds content,
   he is everywhere admired. A model son.

Choiseul,
   surly chauffeur from Clauzel's garage,
   bangs Troy's gate shut!
   It hinges on a scream. His rusty
   common-law wife's. Hands hard as a crank handle,
   he is obsequious, in love with engines.
   They can be reconstructed. Before
   human complications, his horny hands are thumbs.
   Now, seal your eyes, and think of Homer's grief.

Darnley,
> skin freckled like a mango leaf,
> feels the sun's fingers press his lids.
> His half-brother Russell steers him by the hand.
> Seeing him, I practise blindness.
> Homer and Milton in their owl-blind towers,
> I envy him his great affliction. Sunlight
> whitens him like a negative.

Emanuel
> Auguste, out in the harbour, lone Odysseus,
> tattooed ex-merchant sailor, rows alone
> through the rosebloom of dawn to chuckling oars
> measured, dip, pentametrical, reciting
> through narrowed eyes as his blades scissor silk,

>> "Ah moon / (bend, stroke)
>> of my delight / (bend, stroke)
>> that knows no wane.
>> The moon of heaven / (bend, stroke)
>> is rising once again,"

> defiling past Troy town, his rented oars
> remembering what seas, what smoking shores?

FARAH & RAWLINS, temple with
> plate-glass front, gutted, but girded by
> Ionic columns, before which mincing

Gaga,
> the town's transvestite, housemaid's darling,

        is window-shopping, swirling his plastic bag,
        before his houseboy's roundtrip to Barbados,
        most Greek of all, the love that hath no name, and

Helen?
        Janie, the town's one clear-complexioned whore,
        with two tow-headed children in her tow,
        she sleeps with sailors only, her black
        hair electrical
        as all that trouble over Troy,
        rolling broad-beamed she leaves
        a plump and pumping vacancy,
        "O promise me," as in her satin sea-heave follow
        cries of

Ityn! Tin! Tin!
        from Philomène, the bird-brained idiot girl,
        eyes skittering as the sea-swallow
        since her rape,
        laying on lust, in her unspeakable tongue,
        her silent curse.

Joumard,
        the fowl-thief with his cockerel's strut,
        heads home like Jason, in his fluttering coat
        a smoke-drugged guinea hen,
        the golden fleece,

Kyrie! kyrie! twitter
        a choir of surpliced blackbirds in the pews
        of telephone wires, bringing day to

Ligier,
> reprieved murderer, tangled in his pipe smoke,
> wrestling Laocoön,
> bringing more gold to

Midas,
> Monsieur Auguste Manoir,
> pillar of business and the Church,
> rising to watch the sunlight work for him,
> gilding the wharf's warehouses with his name.

Nessus,
> nicknamed N'homme Maman Migrain
> (your louse's mother's man),
> rises in sackcloth, prophesying
> fire and brimstone on the gilt wooden towers of
> offices, ordures, on
> Peter & Co. to burn like Pompeii, on J.
> Q. Charles's stores, on the teetering, scabrous City of
> Refuge, my old grandmother's barracks, where, once

Submarine,
> the seven-foot-high bum-boatman,
> loose, lank, and gangling as a frayed cheroot,
> once asking to see a ship's captain, and refused,
> with infinite courtesy bending, inquired,
> "So what the hell is your captain?
> A fucking microbe?"

Troy town awakens,
> in its shirt of fire, but on our street

Uncle Eric
    sits in a shadowed corner,
    mumbling, hum-eyed,
    writing his letters to the world,
    his tilted hand scrambling for foothold.

Vaughan,
    battling his itch, waits for the rumshop's
    New Jerusalem, while Mister

Weekes,
    slippered black grocer in gold-rimmed spectacles,
    paddles across a rug of yellow sunshine
    laid at his feet by the shadows of tall houses,
    towards his dark shop,
    propelled in his tranced passage by one star:
    Garvey's imperial emblem of Africa United,
    felt slippers muttering in Barbadian brogue,
    and, entering his shop,
    is mantled like a cleric
    in a soutane of onion smells, saltfish, and garlic,
    salt-flaked Newfoundland cod hacked by a cleaver
    on a scarred counter where a bent halfpenny
    shows Edward VII, Defender of the Faith, Emperor of India,
    next to a Lincoln penny, IN GOD WE TRUST
    "and in God one, b' Christ," thinks Mr. Weekes,
    opening his Bible near the paradise plums,
    arm crooked all day over a window open
    at the New Jerusalem, for Coloured People Only.
    At Exodus.

Xodus, bearing back the saxophonist,
    Yes, whose ramshorn is his dented saxophone,
    bearing back to the green grasses of Guinea,

Zandoli,
    nicknamed The Lizard,
    rodent exterminator, mosquito murderer,
    equipment slung over a phthisic shoulder,
    safariing from Mary Ann Street's café,
    wiping a gum-bright grin, out for the week's assault on
    roaches, midges, jiggers, rodents, bugs, and larvae,
    singing, refumigating
    Jerusalem, for Coloured People Only.

These dead, these derelicts,
that alphabet of the emaciated,
they were the stars of my mythology.

Chapter 4

> *—Jerusalem, the golden,*
> *With milk and honey blest*

Thin water glazed
                the pebbled knuckles of the Baptist's feet.
In Craven's book.
Their haloes shone like the tin guards of lamps.
Verocchio. Leonardo painted the kneeling angel's hair.
Kneeling in our plain chapel,
I envied them their frescoes.
Italy flung round my shoulders like a robe,
I ran among dry rocks, howling, "Repent!"
Zinnias, or else some coarser marigold,
brazenly rigid in their metal bowls
or our porch's allamandas trumpeted
from the Vermeer white napery of the altar:
LET US COME INTO HIS PRESENCE WITH THANKSGIVING
AND INTO HIS COURTS WITH PRAISE.
                              Those bowls,
in whose bossed brass the stewards were repeated
and multiplied, as in an insect's eye,
some jewelled insect in a corner of Crivelli,
were often ours, as were the trumpet flowers
between the silvered chargers with the Host
and ruby blood.

Collect, epistle, lesson,
the Jacobean English rang, new-minted,
the speech of simple men,
evangelists, reformers, abolitionists,
their text was cold brook water,
they fell to foreign fevers,
I would be a preacher,
I would write great hymns.

Arnold, staid melancholy of those Sabbath dusks,
I know those rigorous teachers of your youth,
Victorian gravures of the Holy Land,
thorn-tortured Palestine,
bearded disciples wrapped tight in malaria,
the light of desert fevers,
and those thin sunsets
with the consistency of pumpkin soup.
Grey chapel where parched and fiery Reverend Pilgrims
were shrieking twigs,
frock-coated beetles gesturing hell-fire.
Are you cast down, cast down, my coal-black kin?
Be not afraid, the Lord shall raise you up.

The cloven hoof, the hairy paw
despite the passionate, pragmatic
Methodism of my infancy,
crawled through the thicket of my hair,
till sometimes the skin prickled
even in sunshine at "negromancy";
traumatic, tribal,
an atavism stronger than their Mass,
stronger than chapel, whose

tubers gripped the rooted middle class,
beginning where Africa began:
in the body's memory.

I knew them all,
the "swell-foot," the epileptic *"mal-cadi,"*
cured by stinking compounds,
tisane, bush-bath, the exhausting emetic,
and when these failed, the incurably sored and sick
brought in a litter to the obeah-man.
One step beyond the city was the bush.
One step behind the church door stood the devil.

THE PACT

One daybreak, as the iron light,
which guards dawn like a shopfront, lifted,
a scavenger washing the gutters stood
dumbstruck at the cross
where Chaussee Road and Grass Street intersect
before a rusting bloodstain.
A bubbling font at which
a synod of parsonical flies presided,
washing their hands. The scavenger Zandoli
slowly crossed himself.
The slowly sinking stain mapped no direction
in which the thing, a dog perhaps, had crawled.
Light flushed its crimson like an obscene rose.
A knot of black communicants,
mainly old women, chorused round the wound.
The asphalt, like an artery, flowed, unstanched.

Monsieur Auguste Manoir, pillar of the Church,
lay on his back and watched dawn ring
his bed's gold quoits, and gild the view
of hills and roofs the hue of crusted blood,
heard the grey, iron harbour
open on a seagull's rusty hinge,
and knew, as soundlessly as sunlight,
that today he would die.

The blood of garbage mongrels had a thin,
watery excretion; this, a rich red
bubbled before their eyes.

Monsieur Manoir
urged his ringed, hairy hand to climb his stomach
to nuzzle at his heart.
Its crabbed jaw clenched the crucifix;
he heaved there, wheezing,
in the pose of one swearing eternal fealty,
hearing his blood race
like wine from a barrel when its bung has burst.

The blood coagulated like dregged wine.
Zandoli hefted a bucket
washing it wide. It
spread like a dying crab, clenching the earth.
Laved in a sudden wash of sweat, Manoir
struggled to scream for help.
His wife, in black, bent at communion.
Released, he watched the light deliriously dancing
on the cold, iron roofs of his warehouses
whose corrugations rippled with his name.

His hands still smelled of fish, of his beginnings,
hands that he'd ringed with gold, to hide their smell;
sometimes he'd hold them out,
puckered with lotions, powdered, to his wife,
a peasant's hands, a butcher's,
their acrid odour of saltfish and lard.
Drawn by the sweat,
a fly prayed at his ear well:

Bon Dieu, pardon,
Demou, merci,
l'odeur savon,
l'odeur parfum
pas sait guérir
l'odeur péché,
l'odeur d'enfer,
pardonnez-moi
Auguste Manoir!

If there was one thing Manoir's watchman hated
more than the merchant, it was the merchant's dog,
more wolf than dog. It would break loose
some nights, rooting at the warehouse,
paws scuffing dirt like hands for some lost bone.
Before he struck it, something dimmed its eyes.
Its head dilating like an obscene rose,
humming and gemmed with flies, the dog
tottered through the tiled hallway of the house
towards its bed.

Under a scabrous roof whose fences
held the colours of dried blood, Saylie,

the wrinkled washerwoman, howled
in gibberish, in the devil's Latin.
Stepping back from the stench
as powerful as a cloud of smoke
the young priest chanted:
per factotem mundi,
per eum qui habet potestatem
mittendi te in gehennam . . .
six men with difficulty pinning her down,
gasping like divers coming up for breath,
her wild eyes rocketing,
as Beherit and Eazaz wrestled in her smoke.

The stores opened for business.
A stench of rumour filtered through the streets.
He was the first black merchant baron.
They would say things, of course, they would think things,
those children of his fellow villagers
descending the serpentine roads from the Morne,
they'd say his name in whispers now, "Manoir."

The priest prayed swiftly, averting his head;
she had, he knew, contracted with the devil,
now, dying, his dog's teeth tugged at her soul
like cloth in a wringer when the cogs have caught,
their hands pulled at its stuff through her clenched teeth,
"Name him!" The priest intoned, "Name! *Déparlez!*"
The bloodstain in the street dried quick as sweat.
"Manoir," she screamed, "that dog, Auguste Manoir."

Chapter 5

—Statio haud malefida carinis

I

AUGUSTE MANOIR, MERCHANT: LICENSED TO SELL
INTOXICATING LIQUOR, RETAILER, DRY GOODS, etc.
his signs peppered the wharves.
From the canted barracks of the City of Refuge,
from his grandmother's tea shop, he would watch
on black hills of imported anthracite
the frieze of coal-black carriers, *charbonniers*,
erect, repetitive as hieroglyphs
descending and ascending the steep ramps,
building the pyramids,
songs of Egyptian bondage,
                                  when they sang,
the burden of the panniered anthracite,
one hundredweight to every woman
tautened, like cable, the hawsers in their necks.
There was disease inhaled in the coal dust.
Silicosis. Herring gulls
white as the uniforms of tally clerks,
screeching, numbered and tagged the loads.

"Boy! Name the great harbours of the world!"
"Sydney! Sir."

"San Fransceesco!"
"Naples, sah!"
"And what about Castries?"
"Sah, Castries ees a coaling station and
der twenty-seventh best harba in der worl'!
In eet the entire Breetesh Navy can be heeden!"
"What is the motto of St. Lucia, boy?"
"*Statio haud malefida carinis.*"
"Sir!"
"Sir!"
"And what does that mean?"
"Sir, a safe anchorage for sheeps!"

High on the Morne,
flowers medalled the gravestones of the Inniskillings,
too late. Bamboos burst like funereal gunfire.
Noon smoke of cannon fodder,
as black bat cries recited Vergil's tag: *"Statio haud!"*
Safe in their anchorage, sloe-lidded sloops
admired their reflections: *Phyllis Mark,
Albertha Compton, Lady Joy, The Jewel.*

II

The teetering two-storeyed house next door became a haven
for bat-like transients.
Tenants flashed in and out of its dark rooms.
Their cries shot from its eaves. A family of Creoles.
The mother a yellow, formidable Martiniquaise,
handsome, obliquely masculine, with a mole, *"très égyptienne,"*
black sapodilla-seed eyes

under the ziggurat of her pompadour,
we called the Captain's Wife.

Sometimes, when the wind's hand creaked her upstairs window or
hiding in the dim angle of our bedroom
I'd try to catch her naked. Their son, Gentile,
had round, scared eyes, a mouth
that gibbered in perpetual terror,
even in sunshine he shivered like a foundling.
"Gentile, Gentile!" we called. His own name frightened him.

We all knew when the Captain had dry-docked.
There would be violent bursts of shrieking French,
and in my own bed, parallel, separated by a gulf
of air, I'd hear the Captain's Wife,
sobbing, denying.
Next day her golden face seemed shrunken,
then, when he ulysseed, she bloomed again,
the bat-swift transients returned,
so many, perhaps they quartered in the eaves.
Dressed in black lace, like an impatient widow,
I imagined that skin, pomegranate, under silks
the sheen of water, and that
sweet-sour stink vixens give off.

Serene, and unimaginably naked,
as her dark countrymen hung round her rooms,
we heard their laughter tinkling above the glasses.
They came when Foquarde travelled down the coast.
Her laugh rang like the jangling of bracelets.

III

Jewel, a single-stack, diesel, forty-foot coastal vessel,
its cabin curtained with canvas meant to shield
passengers from the sun,
but through which rain and shining spray still drenched,
coughed like a relic out of Conrad. Twice a week
she loaded her cargo of pigs, charcoal, food, lumber,
squabbling or frightened peasants, the odd priest,
threading the island's jettied villages,
Anse La Raye, Canaries, Soufrière, Choiseul,
and back. She also carried mail.

In deepgreen village coves she rocked offshore,
threatening her breakdown,
while rust bled from her wash,
a litter of dugouts nuzzling at her flank,
off-loading goods and passengers.

Disembarkation was precarious,
the inshore swell had to be nicely timed
against the lunge of struggling canoes
in which, feet planted squarely as a mast, one man
stood, swaying, heaving with the swell.

IV

Her course sheered perilously close to the ochre rocks
and bushy outcrops of the leeward coast,
sometimes so closely that it seemed to us
"that all the shoreline's leaves were magnified

deliberately, with frightening detail,"
yet the yellow coast uncoiling past her prow
like new rope from a bollard never lost interest,
especially when the coiled beach lay
between black coves blinding
half-moon of sand,
before some settlement which the passengers,
however often they had made this journey,
always gave different names,
"because it went on repeating itself exactly,"
palms, naked children fishing, wretched huts,
a stone church by a brown, clogged river,
the leper colony of Malgrétoute.

A church, hedged by an unconverted forest,
a beach without a footprint, clear or malformed,
no children, no one, on the hollow pier.
The Jewel hove to, ringing her leper's bell.
The passengers crossed themselves and turned,
inevitably, to the priest.
He'd rise as the canoe appeared.
Condemned. I searched his skin.
The surpliced water heaved.
The bell tinkled like Mass. The priest got off.

He sat still in the long canoe, the afternoon
swallowed the bell rings slowly,
one hand steadying his hat,
the other gripping its stern.
After a while we lost him to the dark green
ocean of the leaves, a white speck, a sail,
out of our memory and our gratitude.

Chapter 6

I

Surpliced, processional,
the shallows mutter in Latin,
maris stella, maris stella,

lichens of leprosy,
disconsolate plumes
of the cabbage palms' casques,

alleluia!

Oversexed cockerel,
cutthroat of dawn,
rattle your wattle,

gloria!

A mast tied with flowers
marries canoe and river,
ora!

The bell-mouth of twilight
chafes like a sore,

serene eye of blue
blind, flaking on rock-prow,

foam-snooded figurehead
with her foam-plume lily,
maris stella.

Maria, Maria,
your bows nod benediction,
the broken pier kneels,

sanctus, sanctus,
from the tonsured mountains
the slow stink of incense

from Soufrière's chancre
the volcano's
sulphurous censer,

Sancta Lucia,
an island brittle
as a Lenten biscuit,

a map of cracked precipices
un pain d'épice
christened by Vergil,

*statio haud
malefida carinis,*
screaking from pulleys

the sail's abrupt sanctus.
Vespers. From chapel
the tinkle of a sheep's bell

draws the sea-flocks homeward.

   II

At Canaries,
the sea's steel razor
shines.

Broken, decrepit port
for some rum-eyed romantic,
his empire's secret rusting in a sea-chest,

tarred, tattered coconuts,
an exile's niche
for some Tuan Jim, "a water clerk under a cloud"

(*The T.L.S.*).
Roofs of tin,
pray for us.
The one bell bangs. Too loud.

At Anse La Raye,
moving among pot-stomached, dribbling, snotted,
starved, fig-navelled, mud-baked cherubim,
the French priest strolls down to the pier.
The Jewel,

rounding the point, bringing mail,

the unfinished Government road,
"right round the island," will not be finished this year.

A fish plops. Making rings
that marry the wide harbour.
Navel of life. Perhaps he should have married.

His head is a burnt shell.
This was his heaven once. It smells like hell.
"And what is hell, my children?"

Qui côté c'est l'enfer?
Why, Father, on this coast.
Father, hell is

two hundred shacks on wooden stilts,
one bushy path to the night-soil pits.

Hell is this hole where the devil shits,
but tinkle your mission bell,
M.V. *Jewel*, for

Soufrière, where
the raw
sore of the volcano chafes,
exhausted boil.
Maleboge, where some golden Louis
buried his soldiers,
where the green fusiliers boiled themselves like lobsters,
their ranks bled white,
mute clangour now under the metal leaves,
the golden cocoa's tattered epaulettes,

Alouette, gentille alouette,
tune of the Lark's legions gone to hack pasta,
Louis, their sun-king, sinking
like metal coins under the black leaves,
the stink of sulphur unendurable,
like Marat's bath-pit.
That very special reek,
tristes, tristes tropiques.

Under the Pitons, the green
bay, dark as oil.
Breasts of a woman, serenely rising.
Thought Capitaine Foquarde,
"This is good soil."

III

Palm trunks irregular as railings fenced the beaches.
Behind them: private property, the island quartered
into baronial estates, gone, gone,
their golden, bugled epoch.
Aubrey Smith characters in khaki helmets,
Victorian flourish of oratorical moustaches.
Retired Captain X, who kept an open grave behind his house,
would shoot on sight. Shot himself, sah!
B. reputedly galloped his charger through the canes,
pointed his whip at nubile coolie girls, "Up to the house,"
droit de seigneur, keeping employment in the family.
Timid, bespectacled Monsieur D., more bank clerk than sugar baron,
piety on a horse,
and Royer was murdered in dark cocoa bushes, Boscobel,
by three niggers stealing his coconuts.

The air there, stained. Furred. Soft as tarantulas
scuttling the hairy cocoa leaves.
Baron, ship chandler, merchant, water clerk,
the fiction of their own lives claimed each one.
My fiery grandfather,
his house burnt at Choiseul, and he inside it.
Froth of malaria on the pool's lip.
Bilharzia enters the intestines of small children,
a sort of river tsetse, mines, in the guts,
in labourers, producing lethargy. "We cure it,"
said the young research scientist,
"and multiply the unemployment problem."

IV

Heureux qui comme Ulysse,
ou Capitaine Foquarde,
while his pomegranate-skinned
Martiniquan Penelope
rocks in her bentwood chair,
laughing, stitching ripped knickers,
as her coast-threading captain
hums, *"La vie c'est un voyage"*
and the polished rocker dips
as her white burst of laughter
drives deep whose prow?

Pigs bring bad luck. And priests.
The shorn priest sat to starboard,
cowled by torn canvas,
numbering the beads of villages,
Foquarde. Coquarde.

Lost, lost, rain-hidden, precipitous, debased,
ocean's soiled lace around her dirty ankles.
The ship's bell rankled. At the greased wheel
Foquarde turned, looking seaward.

A tanker.
        Her red, dipping prow,
                              a mouth,
remembering the names of islands.

Chapter 7

I

Provincialism loves the pseudo-epic,
so if these heroes have been given a stature
disproportionate to their cramped lives,
remember I beheld them at knee-height,
and that their thunderous exchanges
rumbled like gods about another life,
as now, I hope, some child
ascribes their grandeur to Gregorias.
Remember years must pass before he saw an orchestra,
a train, a theatre, the spark-coloured leaves
of autumn whirling from a rail line,
that, as for the seasons,
the works he read described their passage with
processional arrogance; then pardon, life,
if he saw autumn in a rusted leaf.
What else was he but a divided child?

I saw, as through the glass of some provincial gallery
the hieratic objects which my father loved:
the stuffed dark nightingale of Keats,
bead-eyed, snow-headed eagles,
all that romantic taxidermy,
and each one was a fragment of the True Cross,
each one upheld, as if it were The Host;

those venerated, venerable objects
borne by the black hands (reflecting like mahogany)
of reverential teachers, shone the more
they were repolished by our use.

The Church upheld the Word, but this new Word
was here, attainable
to my own hand,
in the deep country it found the natural man,
generous, rooted.
And I now yearned to suffer for that life,
I looked for some ancestral, tribal country,
I heard its clear tongue over the clean stones
of the river, I looked from the bus window
and multiplied the bush with savages,
speckled the leaves with jaguar and deer,
I changed those crusted boulders
to grey, stone-lidded crocodiles,
my head shrieked with metallic, raucous parrots,
I held my breath as savages grinned,
stalking, through the bush.

II

About the August of my fourteenth year
I lost my self somewhere above a valley
owned by a spinster-farmer, my dead father's friend.
At the hill's edge there was a scarp
with bushes and boulders stuck in its side.
Afternoon light ripened the valley,
rifling smoke climbed from small labourers' houses,
and I dissolved into a trance.

I was seized by a pity more profound
than my young body could bear, I climbed
with the labouring smoke,
I drowned in labouring breakers of bright cloud,
then uncontrollably I began to weep,
inwardly, without tears, with a serene extinction
of all sense; I felt compelled to kneel,
I wept for nothing and for everything,
I wept for the earth of the hill under my knees,
for the grass, the pebbles, for the cooking smoke
above the labourers' houses like a cry,
for unheard avalanches of white cloud,
but "darker grows the valley, more and more forgetting."
For their lights still shine through the hovels like litmus,
the smoking lamp still slowly says its prayer,
the poor still move behind their tinted scrim,
the taste of water is still shared everywhere,
but in that ship of night, locked in together,
through which, like chains, a little light might leak,
something still fastens us forever to the poor.

But which was the true light?
Blare noon or twilight,
"the lonely light that Samuel Palmer engraved,"
or the cold
iron entering the soul, as the soul sank
out of belief.
          That bugle-coloured twilight
blew the withdrawal not of legions and proconsuls,
but of pale, prebendary clerks, with the gait and gall
of camels. And yet I envied them,
bent, silent decipherers of sacred texts,

their Roman arches, Vergilian terraces,
their tiered, ordered colonial world
where evening, like the talons of a bird
bent the blue jacaranda softly, and smoke rose with
the leisure and frailty of recollection,
I learnt their strict necrology of dead kings,
bones freckling the rushes of damp tombs,
the light-furred luminous world of Claude,
their ruined temples, and in drizzling twilights, Turner.

    III

Our father,
        who floated in the vaults of Michelangelo,
St. Raphael,
        of sienna and gold leaf,
it was then
        that he fell in love, having no care
for truth,
        that he could enter the doorway of a triptych,
that he believed
        those three stiff horsemen cantering past a rock,
        towards jewelled cities on a cracked horizon,
        that the lances of Uccello shivered him,
        like Saul, unhorsed,
that he fell in love with art,
        and life began.

    IV

Noon,
        and its sacred water sprinkles.

A schoolgirl in blue and white uniform,
her golden plaits a simple coronet
out of Angelico, a fine sweat on her forehead,
hair where the twilight singed and signed its epoch.
And a young man going home.
They move away from each other.
They are moving towards each other.
His head roars with hunger and poems.
His hand is trembling to recite her name.
She clutches her books, she is laughing,
her uniformed companions laughing.
She laughs till she is near tears.

      V

Who could tell, in "the crossing of that pair"
      that later it would mean
that rigid iron lines were drawn between
      him and that garden chair
from which she rose to greet him, as for a train,
      that watching her rise
from the bright boathouse door was like some station
      where either stood, transfixed
by the rattling telegraph of carriage windows
      flashing goodbyes,
that every dusk rehearsed a separation
      already in their eyes,
that later, when they sat in silence, seaward,
      and, looking upward, heard
its engines as some moonlit liner chirred
      from the black harbour outward,
those lights spelt out their sentence, word by word?

Two

HOMAGE TO GREGORIAS

I saw them growing gaunt and pale in their unlighted studios. The Indian turning green, the Negro's smile gone, the white man more perverted—more and more forgetful of the sun they had left behind, trying desperately to imitate what came naturally to those whose rightful place was in the net. Years later, having frittered away their youth, they would return with vacant eyes, all initiative gone, without heart to set themselves the only task appropriate to the milieu that was slowly revealing to me the nature of its values: Adam's task of giving things their names.

ALEJO CARPENTIER
The Lost Steps

Chapter 8

—*West Indian Gothic*

I

A gaunt, gabled house,
grey, fretted, soars
above a verdigris canal which
sours with moss. A bridge,
lithe as a schoolboy's leap,
vaults the canal. Each
longitudinal window seems
a vertical sarcophagus, a niche
in which its family must sleep
erect, repetitive as saints
in their cathedral crypt,
like urgent angels in their fluted stone
sailing their stone dream.
And like their house,
all the Gregoriases
were pious, arrogant men,
of that first afternoon, when
Gregorias ushered me in there,
I recall an air of bugled orders,
cavalry charges of children
tumbling down the stair,
a bristling, courteous father,

but also something delicate,
a dessicating frailty which showed
in his worn mother, a taut tree
shorn to the dark house's use,
its hothouse, fragile atmosphere
labouring yearly to produce
the specimen *Gregorias elongatus*.

In the spear-lowering light of afternoon
I paced his hunter's stride,
there was a hierarchic arrogance in his bearing
which crested in the martial,
oracular moustaches of his father,
a Lewis gunner in the First World War,
now brown, prehensile fingers plucked his work,
lurid Madonnas, pietistic crucifixions
modelled on common Catholic lithographs,
but with the personal flourish of a witness.
Widowed, his father's interest in life declined,
his battle finished. The brown twigs broke apart.

Around that golden year which I described
Gregorias and that finished soldier quartered
in a brown, broken-down bungalow
whose yard was indistinguishable from bush,
between the broad-leaved jungle and the town.
Shaky, half-rotted treaders, sighing, climbed
towards a sun-warped verandah, one half of which
Gregorias had screened into a studio,
shading a varnished, three-legged table
crawling with exhausted paint tubes, a lowering quart
of Pirate rum, and grey, dog-eared, turpentine-stained editions

of the Old Masters. One day the floor collapsed.
The old soldier sank suddenly to his waist,
wearing the verandah like a belt.
Gregorias buckled with laughter telling this,
but shame broke the old warrior.
The dusk lowered his lances through the leaves.
In another year the soldier shrank and died.
Embittered, Gregorias wanted carved on his stone:
PRAISE YOUR GOD, DRINK YOUR RUM, MIND YOUR OWN BUSINESS.

We were both fatherless now, and often drunk.

Drunk,
        on a half pint of joiner's turpentine,
drunk,
        while the black, black-sweatered, horn-soled fishermen drank
        their *l'absinthe* in sand back yards standing up,
        on the clear beer of sunrise,
        on cheap, tannic Canaries muscatel,
        on glue, on linseed oil, on kerosene,
        as Van Gogh's shadow rippling on a cornfield,
        on Cézanne's boots grinding the stones of Aix
        to shales of slate, ochre, and Vigie blue,
        on Gauguin's hand shaking the gin-coloured dew
        from the umbrella yams,
        garrulous, all day, sun-struck,
till dusk glazed vision with its darkening varnish.
Days welded by the sun's torch into days!
Gregorias plunging whole-suit in the shallows,
painting under water, roaring, and spewing spray,
Gregorias gesturing, under the coconuts'
wickerwork shade–tin glare–wickerwork shade,

days woven into days, a stinging haze
of thorn trees bent like green flames by the Trades,
under a sky tacked to the horizon, drumskin tight,
as shaggy combers leisurely beard the rocks,
while the asphalt sweats its mirages and the beaks
of fledgling ginger lilies
gasped for rain.
Gregorias, the easel rifled on his shoulder, marching
towards an Atlantic flashing tinfoil,
singing "O Paradiso,"
till the western breakers laboured to that music,
his canvas crucified against a tree.

II

But drunkenly, or secretly, we swore,
disciples of that astigmatic saint,
that we would never leave the island
until we had put down, in paint, in words,
as palmists learn the network of a hand,
all of its sunken, leaf-choked ravines,
every neglected, self-pitying inlet
muttering in brackish dialect, the ropes of mangroves
from which old soldier crabs slipped
surrendering to slush,
each ochre track seeking some hilltop and
losing itself in an unfinished phrase,
under sand shipyards where the burnt-out palms
inverted the design of unrigged schooners,
entering forests, boiling with life,
goyave, corrosol, bois-canot, sapotille.

Days!
The sun drumming, drumming,
past the defeated pennons of the palms,
roads limp from sunstroke,
past green flutes of grass
the ocean cannonading, come!
Wonder that opened like the fan
of the dividing fronds
on some noon-struck Sahara,
where my heart from its rib cage yelped like a pup
after clouds of sanderlings rustily wheeling
the world on its ancient,
invisible axis,
the breakers slow-dolphining over more breakers,
to swivel our easels down, as firm
as conquerors who had discovered home.

III

For no one had yet written of this landscape
that it was possible, though there were sounds
given to its varieties of wood;

the *bois-canot* responded to its echo,
when the axe spoke, weeds ran up to the knee
like bastard children, hiding in their names,

whole generations died, unchristened,
growths hidden in green darkness, forests
of history thickening with amnesia,

so that a man's branched, naked trunk,
its roots crusted with dirt,
swayed where it stopped, remembering another name;

breaking a lime leaf,
cracking an acrid ginger root,
a smell of tribal medicine stained the mind,

stronger than ocean's rags,
than the reek of the maingot forbidden pregnant women,
than the smell of the horizon's rusting rim,

here was a life older than geography,
as the leaves of edible roots opened their pages
at the child's last lesson, Africa, heart-shaped,

and the lost Arawak hieroglyphs and signs
were razed from slates by sponges of the rain,
their symbols mixed with lichen,

the archipelago like a broken root,
divided among tribes, while trees and men
laboured assiduously, silently to become

whatever their given sounds resembled,
ironwood, logwood-heart, golden apples, cedars,
and were nearly

ironwood, logwood-heart, golden apples, cedars,
men . . .

Chapter 9

I

There are already, invisible on canvas,
lines locking into outlines. The visible dissolves
in a benign acid. The leaf
insists on its oval echo, that wall
breaks into sweat, oil settles
in the twin pans of the eyes.

Blue, on the tip of the tongue,
and this cloud can go no further.
Over your shoulder the landscape
frowns at its image. A rigour
of zinc white seizes the wall,
April ignites the immortelle,
the leaf of a kneeling sapling
is the yellow flame of Lippi's "Annunciation."
Like the scrape of a struck match, cadmium orange,
evened to the wick of a lantern.
Like a crowd, surrounding the frame,
the muttering variegations of green.

The mountain's crouching back begins to ache.

The eyes sweat, small fires gnaw
at the edge of the canvas,

ochre, sienna, their smoke
billows into blue cloud.
A bird's cry tries to pierce
the thick silence of canvas.
From the reeds of your lashes, the wild commas
of crows are beginning to rise.
At your feet
the dead cricket grows into a dragon,
the razor grass bristles resentment,
gnats are sawing the air,
the sun plates your back,
salt singes your eyes
and a crab, the brush in its pincer,
scrapes the white sand of canvas,

as the sea's huge eye stuns you
with the lumbering, oblique blow
of its weary, pelagic eyelid,
its jaw ruminates
on the seagrass it munches
while the lighthouse needle signals
like a stuttering compass
north north by northwest north
and your hair roars like an oven
and a cloud passes,
till the landscape settles on
a horizon humming with balance,
and like a tired sitter
the world shifts its weight.

Remember Vincent, saint
of all sunstroke, remember

Paul, their heads
plated with fire!
The sun explodes into irises,
the shadows are crossing like crows,
they settle, clawing the hair,
yellow is screaming.

Dear Theo, I shall go mad.

Is that where it lies,
in the light of that leaf, the glint
of some gully, in a day
glinting with mica, in that rock
that shatters in slate,
in that flashing buckle of ocean?
The skull is sucked dry as a seed,
the landscape is finished.
The ants blacken it, signing.
Round the roar of an oven, the gnats
hiss their finical contradiction.
Nature is a fire,
through the door of this landscape
I have entered a furnace.

I rise, ringing with sunstroke!

The foreground lurches up drunkenly,
the cold sea is coiled in your gut,
the sky's ring dilates, dilates, and
the tongue tastes sand,
the mouth is sour with failure,
the hair on your nape,

spiders running over your wrist,
stirs like trees on the edge of that ridge,
you have eaten nothing but this landscape
all day, from daybreak to noon and past noon
the acrid greens and ochres
rust in the gut.
The stomach heaves, look away.
Your lashes settle like crows.

I have toiled all of life for this failure.
Beyond this frame, deceptive, indifferent,
nature returns to its work,
behind the square of blue you have cut from that sky,
another life, real, indifferent, resumes.
Let the hole heal itself.
The window is shut.
The eyelids cool in the shade.
Nothing will show after this, nothing
except the frame which you carry in your sealed, surrendering eyes.

II

Where did I fail? I could draw,
I was disciplined, humble, I rendered
the visible world that I saw
exactly, yet it hindered me, for
in every surface I sought
the paradoxical flash of an instant
in which every facet was caught
in a crystal of ambiguities,
I hoped that both disciplines might
by painful accretion cohere

and finally ignite,
but I lived in a different gift,
its element metaphor,
while Gregorias would draw
with the linear elation of an eel
one muscle in one thought,
my hand was crabbed by that style,
this epoch, that school
or the next, it shared
the translucent soul of the fish, while
Gregorias abandoned apprenticeship
to the errors of his own soul,
it was classic versus romantic
perhaps, it was water and fire,
and how often my hand betrayed
creeping across the white sand,
poor crab, its circuitous instinct
to fasten on what it seized,
but I was his runner, I paced him,
I admired the explosion of impulse,
I envied and understood
his mountainous derision
at this sidewise crawling, this classic
condition of servitude.
His work was grotesque, but whole,
and however bad it became
it was his, he possessed
aboriginal force and it came
as the carver comes out of the wood.
Now, every landscape we entered
was already signed with his name.

III

In my father's small blue library
of reproductions I would find
that fine-drawn hare of Dürer's, clenched and quivering
to leap across my wrist,
and volumes of *The English Topographical Draughtsmen*,
Peter de Wint, Paul Sandby, Cotman, and in another
sky-blue book
the shepherdesses of Boucher and Fragonard,
and I raved for
the split pears of their arses,
their milk-jug bubs,
the close and, I guessed, golden
inlay of curls at cunt
and conch-like ear,
and after service, Sunday lay
golden, a fucked Eve
replete and apple-bearing,
and if they were my Muse,
still, out of that you rose,
body downed with the seasons,
gold and white, Anna
of the peach-furred body, light
of another epoch,
and stone-grey eyes.
Was the love wrong that came
out of the Book of Hours,
and the reaper with his scythe,
as your hair gold, dress green,
sickle-armed, you move

through a frontispiece of flowers
eternal, true as Ruth
the wheat sheaves at her ear,
or gorgonizing Judith
swinging the dead lantern
of Holofernes, that bright year
like all first love, we were
pure and Pre-Raphaelite,
Circe-coil of plaited light
around her, as Gregorias
bent to his handful of earth,
his black nudes gleaming sweat,
in the tiger shade of the fronds.

Through the swift year
the canvases multiply,
brown-bottomed tumbling cherubim,
broad-bladed breadfruit leaves
surround his oval virgin
under her ringing sky,
the primal vegetation
the mute clangour of lilies,
every brushstroke a prayer
to Giotto, to Masaccio,
his primitive, companionable saints.

Never such faith again, never such innocence!

Chapter 10

—Frescoes of the New World

I

The Roman Catholic church at Gros Ilet,
a fat, cream-coloured hunk of masonry,
bluntly concludes the main street of this village
of kneeling shacks, their fences
burdened with violet rosaries, their hinges
rusted from sea-blast.

                The beach
is plagued with flies, at the lime edge
of channel, their black hymn
drones in the ear. The broken paths debouch
onto a broken pier. Before the church
there is a hedge of altar lace,
a foam-flecked garden whose barefooted path
circles the gabled presbytery
towards a kind of meadow, where a cemetery
rings with the sea. The church is a shell
of tireless silence. The cold vault
cowls like a benediction. There, a young priest,
a Frenchman, Gregorias's friend,
gave him his first commission.

                    Here was his heaven!
Here was the promise, clear, articulated,
here a new paradise of sea-salt, rum, and paint,
sea-wind tempered with turpentine,
his *mundo nuevo,* our new Raphael.
Above the altar lace
he mounted a triptych of the Assumption
with coarse, purpureal clouds, a prescient Madonna
drawn from Leonardo's "Our Lady of the Rocks."
He soared on his trestles, the curled days shorn
by the adze of St. Joseph the Worker,
till dusk, the tree of heaven, broke in gold leaf.

Not salt, but his own sorrow sprinkled those
hard-knuckled yards where thorn
trees starved
like their bent nurturers, who fed
on the priest's sprinkling hand, whose coarse
mud-crusted feet were yams, whose loss
grew its own roots; O that his art
could sink in earth, as fragrant as Christ's tomb!
But darkness hid
never departing wholly as it promised
between the yam leaves on the riverbank
on whose bent heads the rain splintered like mercury,
in shacks like paper lanterns, in green lagoons
whose fading eye held Eden like a transfer,
it hid in yellowing coconuts where the sea-swifts
breaking and forming in their corolla
fluttered like midges round the glare
of the gas lantern. Shadows left the wall, bats
ferried his thoughts across the feverish creek

where the kingfisher startled like a match,
and rows of trees like savages stepped back
from the gas lantern's radiance, from his faith.

Night—and ocean's hiss
enters the lantern. His mind
beds in its nest of snakes.
Over the village the dim stars smelt of sulphur.
They peered through cloth like children with weak eyes
at his life passing.
Like any priest he knew
the darkness of that life was gathering
like oil in the cruses of husked shells,
and that his gift, not he, had made its choice.

II

Gregorias laughs, a white roar ringed with lamplight,
gigantic moths, the shadows of his hands
fluttering the wall, it is his usual
gesture now, the crucifix.
"Man I ent care if they misunderstand me,
I drink my rum, I praise my God, I mind my business!
The thing is you love death and I love life."
Then, wiping the oracular moustache:
"Your poetry too full of spiders,
bones, worms, ants, things eating up each other,
I can't read it. Look!"
He frames a seascape in a chair,
then, striding back, beyond a table littered
with broken loaves, fishbones, a gut-rusting wine,
he smites his forehead.

"Ah, Gregorias, you are a genius, yes!
Yes, God and me, we understand each other."
He hoists his youngest seascape like a child,
kisses, cradles it, opens the window
of the village night, head tilted seaward,
grey gaze serenely clamped,
lean fingers waving, "Listen!"
As if the thunderous Atlantic
were a record he had just put on.
"Listen! Vasco da Gama kneels to the New World."

O Paradiso! sang
the pied shoal at the edge of Half-Moon Battery
netted in sunlight,
swayed in the wash
as the spent swell broke backward.

And it all sang,
surpliced, processional,
the waves clapped their hands, hallelujah!
and the hills were joyful together,
arpeggios of lizards scuttled the leaves,
swift notes, and under earth
the stifled overtures of cannon thunder.

The promontory was hived
with history, tunnels, moss-grimed abandoned armouries.
Across the blue harbour, past the hospital,
a parched and stubbled headland
like a lion's knuckled paw of tawny grass,
lay the insane asylum.
A peel of lemon sand

curled like a rind across the bay's blue dish.
He had his madness,
mine was our history.

III

O oceanic past,
we were like children
emptying the Atlantic with an enamel cup.
I crouched under each crest
the sneering wave,
the stately foam's *perruque*,
I prayed for it to go
under the fisherman's heel,
for the clods to break
in epochs, crumbling Albion
in each unshelving scarp,
Sidon and Tyre,
an undersea museum,
I saw in the glazed, rocking shallows
the sea-wrack of submerged Byzantium,
as the eddies pushed their garbage to this shore.

Crouched there,
like a whelk-picker,
I searched the sea-wrack for a sea-coin:
my white grandfather's face,
I heard in the black howl of cannon,
sea-agape,
my black grandfather's voice,
and envied mad, divine Gregorias
imprisoned in his choice.

IV

But I tired of your whining, Grandfather,
in the whispers of marsh grass,
I tired of your groans, Grandfather,
in the deep ground bass of the combers,
I cursed what the elm remembers,
I hoped for your sea-voices
to hiss from my hand,
for the sea to erase
those names a thin,
tortured child, kneeling, wrote
on his slate of wet sand.

Chapter 11

I

In one house where I went
on irritable sewing errands for my mother,
there was a large tapestry of, was it Waterloo?
a classically chaotic canvas
of snorting, dappled chargers,
their fetlocks folded under swollen bellies,
their nostrils dragonish, smoking,
the bursting marbled eyes elate with fear.
Their riders were a legion of dragoons
sabre-moustached, canted on stiffened rein,
their arms crooked in a scything sweep,
vaulting a heap of dying,
one in the stance of a reclining Venus,
as casual as Giorgione,
surrendering the standard to a sergeant
bent low and gamy as a polo player,
the whole charge like a pukkha, without blood;
in the tainted, cry-haunted fog,
in the grey, flickering mist,
no mouth of pain,
every chivalric wound
rose-lipped, dandiacal, sweet,
every self-sacrifice perfumed;

my head roared with gold.
I bled for all. I thought it full of glory.

Cramming halfheartedly for the scholarship,
I looked up from my red-jacketed Williamson's
History of the British Empire, towards
the barracks' plumed, imperial hillsides
where cannon bursts of bamboo sprayed the ridge,
riding to Khartoum, Rorke's Drift,
through dervishes of dust,
behind the chevroned jalousies
I butchered fellaheen, thuggees, Mamelukes, wogs.

Up, past the chrome-green, chalet-sheltering Morne,
where somebody, possibly Bryan Edwards, wrote:
"Five governors died of yellow fever in that pavilion,"
were the brick barracks. Graves medalled
by pellets of sheepdung, and
the stone curd of the Carib's pothound. Coral.
Tranced at my desk,
groggy with dates, I leant
across my musket. Redcoat ruminant.

   II

Furred heat prickles this redcoat. Its buttons scorch.
Pellets of salt sweat sting my eyes, once blue,
but which in this green furnace are holes drilled through
by the sun's poker. My bones crack like a torch.
The Fifth stews in its billet behind bounding arches,
the muskets are coolest in their dark, greased armoury,

but our scouts drop like bats. These jungle marches
sickle us, parch us like madmen for a sea
so blue it stains. Downhill a black pig squeals.
Smoke climbs, leisure of recollection.
This forest keeps no wounds, this nature heals
the newest scar, each cloud wraps like a bandage
whatever we enact. What? Chivalry. The fiction
of rusted soldiers fallen on a schoolboy's page.

III

I saw history through the sea-washed eyes
of our choleric, ginger-haired headmaster,
beak like an inflamed hawk's,
a lonely Englishman who loved parades,
sailing, and Conrad's prose.
When the war came the mouths began to bleed,
the white wounds put out tongues.

Nostalgia! Hymns of battles not our own,
on which our fathers looked with the black, iron mouths
of cannon, sea-agape,
to the bugle-coloured light crying from the west,
those dates we piped of redoubt and repulse,
while in our wrists the kettle drums pulsed on
to Khartoum, Lucknow, Cawnpore, Balaclava.
"How strange," said Bill (Carr),
"to find the flag of my regiment,"
where on the razorback ridge
the flag of the Inniskillings every sunset
is hung to bleed for an hour.
A history of ennui, defence, disease,

tumuli of red soldier crabs
that calcified in heaps, their carapaces
freckled with yellow fever,
until the Morne hummed like a hospital,
the gold helmets of dragoons like flowers
tumbling down blue crevasses
in this "Gibraltar of the Gulf of Mexico,"
and on the slate-grey graves more
flowers, medalling their breasts, too late,
fade, "like the white plumes of the Fighting Fifth
who wore the feather without stain."

The leaping Caribs whiten,
in one flash, the instant
the race leapt at Sauteurs,
a cataract! One scream of bounding lace.

IV

I am pounding the faces of gods back into the red clay they
leapt from the mattock of heel after heel, as if heel
after heel were my thumbs that once gouged out as sacred
vessels for women the sockets of eyes, the deaf howl
of their mouths, and I have wept less for them dead than I did
when they leapt from my thumbs into birth, than my
heels which have never hurt horses that now pound them
back into what they should never have sprung from,
staying un-named and un-praised where I found them—
in the god-breeding, god-devouring earth!

We are ground as the hooves of their horses open the wound
of those widening cliffs and the horns of green branches come

lowering past me and the sea's crazed horses the foam
of their whinnying mouths and white mane and the pelting red
pepper of flowers that make my eyes water, yet who am I, under
such thunder, dear gods, under the heels of the thousand
racing towards the exclamation of their single name,
Sauteurs! Their leap into the light? I am no more
than that lithe dreaming runner beside me, my son, the roar
of his heart, and their hearts, I am one with this engine
which is greater than victory, and their pride
with its bounty of pardon, I am one
with the thousand runners who will break on loud sand
at Thermopylae, one wave that now cresting must bear
down the torch of this race, I am all, I am one
who feels as he falls with the thousand now his tendons harden
and the wind god, Hourucan, combing his hair . . .

V

In the child's mind
dead fellaheen were heaped in piles of laundry,
and the converted starched with light and sweetness,
white angels flocking round Gordon's golden palms,
the nodding plumes, I SERVE,
resurrected horsemen choiring from the horizon
from the sepia washes of the *Illustrated London News*,
and the child, like a ribbed mongrel
trailing the fading legions,
singing in his grandfather's company.
Peccavi. I have Sind.

Pronounce it. Peace.
Peace to the Fighting Fifth

that wore the feather without stain,
peace to the flutes of bone-shard
and the harvest scythed by fever,
as twilight lowers its lances
against the breastplates of the bole
in quivering shafts, as
night pivots on a seagull's rusted winch.
Deep in the trees a glowworm army haunts
with haunted eyes, their mouths
as soft as moths crying for their lost countries,
translucent yeomanry.
It was their lost blood that rose in her cheek,
their broken buckles which flashed in her hair.

Chapter 12

I

I watched the vowels curl from the tongue of the carpenter's plane,
resinous, fragrant
labials of our forests,
over the plain wood
the back crouched,
the vine-muscled wrist,
like a man rowing,
sweat-fleck on blond cedar.
Disgruntled Dominic.
I sauntered from his dark shop up the Chaussee.

It was still Paris.
                    The twenties. Montparnasse.
Paris, and now its crepuscule
sets in Pound's eye, now as I watch
this twinkling hoarfrost photograph
of the silvery old man bundled, silent, ice-glint
of frozen fire before the enemy,
faites vos hommages,
as the tongues of shavings coil from the moving pen,
to a Paris of plane trees,
to the peeled ease of Hemingway's early prose,
faites vos hommages,
to the hills stippled with violet
as if they had seen Pissarro.

The smell of our own speech,
the smell of baking bread,
of drizzled asphalt, this
odorous cedar. After the rain
the rinsed shingles shone,
resinous as the smell of country sweat,
of salt-crusted fishermen.
Christ, to shake off the cerecloths,
to stride from the magnetic sphere of legends,
from the gigantic myth.
To change the marble sweat which pebbled
the wave-blow of stone brows
for this sweat-drop on the cedar plank,
for a future without heroes,
to make out of these foresters and fishermen
heraldic men!

Sunset and dawn like manacles chafed his wrist,
no day broke without chains,
bent like a carpenter over the new wood,
a galley slave over his scarred desk,
hours breaking over his head in paper,
even in his sleep, his hands
like lolling oars.
Where else to row, but backward?
Beyond origins, to the whale's wash,
to the epicanthic Arawak's Hewanora,
back to the impeachable pastoral,
praying the salt scales would flake from our eyes
for a horned, sea-snoring island
barnacled with madrepore,
without the shafts of palms stuck in her side.

Santa Lucia,
patroness of Naples, they had put out her eyes,
saint of the blind,
whose vision was miraculously restored.

The sun came through our skins,
and we beheld, at last,
the exact, sudden definition
of our shadow.
Under our grinding heel
the island burst to a crushed
odour of hogplums, acrid, exuding
a memory stronger than madeleines;
a bay became a blow
from a burnished forehead,
the sea's grin stupefied,
painter and poet walked
the hot road, history-less.

Dating like Christmas cards,
the poinsettia bled deep
into the year, so bright
you expected blood drops on the grass,
next was the season
of the swamp tree's fire,
the foothills blue as smoke,
it was our real, our false
spring. Beside the road,
a beautiful, brown Indian
girl in rags. Sheaves
of brown rice held in brown,
brittle hands, watching us
with that earth-deep darkness

in her gaze. She was
the new Persephone,
dazed, ignorant,
waiting to be named.

    II

But we were orphans of the nineteenth century,
sedulous to the morals of a style,
we lived by another light,
Victoria's orphans, bats in the banyan boughs.
Dragonfly, dragonfly
over that gilded river
like teatime afternoons with the Old Masters,
in those long pastoral twilights after the war,
dragonfly, your angry vans of gauze
caught in "the light that Samuel Palmer engraved,"
burnt black in the lamp of Giorgione,
dragonfly, in our ears
sang Baudelaire's exhortation to stay drunk,
sang Gauguin's style, awarded Vincent's ear.

I had entered the house of literature as a houseboy,
filched as the slum child stole,
as the young slave appropriated
those heirlooms temptingly left
with the Victorian homilies of *Noli tangere*.
This is my body. Drink.
This is my wine.
Gregorias squeezing the rum-coloured river
with dead limes. Well,
in the beginning, all
drunkenness is Dionysiac, divine.

III

We drank for all our fathers,
for freedom, as for mine,
how Mama'd praise that angular abstinence,
as if that prim-pursed mouth instantly clicked
at the thought of a quick one,
"Your father never drank,"
self-righteous sphincter!
Starved, burning child,
remember "The Hay Wain"
in your museum, Thomas Craven's book?
The time was coming then
to your parched mind,
in love with amber, with another light
in the unheard, creaking axle,
the marble-coloured horse
and the charnel harvest cart,
in the fire-coloured hole eating the woods,
when on the groaning wagon,
the grown mind would sing
"No cure, no cure."

Yet, Gregorias, lit,
we were the light of the world!

IV

And how could we know then,
damned poet and damned painter,
that we too would resemble

those nervous, inflamed men,
fisherman and joiner,
with their quivering addiction
to alcohol and failure,
who hover in a fiction
of flaming palely at doors
for the rumshop lamp to glare,
with watered eyes, loose collars
and the badge of a bone stud,
their vision branched with blood,
their bodies trees which fed
a fire beyond control,
drinkers who lost their pride
when pride in drink was lost.
We saw, within their eyes,
we thought, an artist's ghost,
but dignified, dignified
through days eaten with shame;
we were burned out that year
with the old sacred flame,
we swore to make drink
and art our finishing school,
join brush and pen and name
to the joiner's strenuous tool.

And then, one night, somewhere,
a single outcry rocketed in air,
the thick tongue of a fallen, drunken lamp
licked at its alcohol ringing the floor,
and with the fierce rush of a furnace door
suddenly opened, history was here.

Three

A SIMPLE FLAME

All have actually parted from the house, but all truly have remained. And it's not the memory of them that remains, but they themselves. Nor is it that they remain in the house but that they continue because of the house. The functions and the acts go from the house by train or by plane or on horseback, walking or crawling. What continues in the house is the organ, the gerundial or circular agent. The steps, the kisses, the pardons, the crimes have gone. What continues in the house is the foot, the lips, the eyes, the heart. Negations and affirmations, good and evil have scattered. What continues in the house is the subject of the act.

CESAR VALLEJO
Poemas Humanos

Chapter 13

I

The whole sky caught. The thick sea heaved like petrol.
The past hissed in a cinder.
They heard the century breaking in half.
Then, towards daybreak, rain
sprinkled the cinders. Clouds
steamed from the broken axle tree.

The sky, vibrating, rippled like sheet iron,
like hairy behemoths, their lungs burnt out, the hills
were hoarse with smoke, everywhere
the retching odours of a tannery.
Clouds curled like burnt-out papers at their edges,
the telephone wires sang from pole to pole
parodying perspective. The wall of heat,
now menacing, now a thief giving ground,
stepped back with every step,
the sea was level with the street.
Every lot of desolation stood
absurdly walled, its toothless breach confessing
perverted bedsprings, heat-stained mattresses,
all of the melancholy, monotonous rubbish
of those who thought their lives strange to their neighbours,
their sins repeated tiredly by the same
picture-frames, papers, blue magnesia bottles,

under arched, amnesiac stairways,
hesitating:

      Soon, on the promontory,
an atmosphere resembling what they had read of war
began its bivouac fires, haphazard tents,
lives casually tangled like unsorted laundry.
Then, like rifle fire, the flutes of smoke,
the first, white flags of washing,
were bravely signalling that some pact
of common desolation had begun.
In the ruins, a population of ragpickers,
bent over stones, deciphering their graves.
Hoses plied the shambles
making the ashes mud.
Here were the broken arches and the vines
ascending leisurely, with the languor of fire.
Your ruined Ilion, your grandfather's pyre.

II

A landscape of burnt stones and broken arches
arranged itself with a baroque panache.
On somnolent, townless twilights, when
piano practice sprinkled the dyed air
before the grid of stars budded with lights
above a system of extinguished streets
he headed for the promontory across the harbour,
from the old town to where in the old barracks
the refugees began another life.
Buckets clanged under the public pipes,
furred dusk clawed softly, mewing at his ear,

all of the sounds of evening fell on velvet,
the night was polishing star after star,
the mild, magnificent night with all its studs on,
buttoned and soldierly, with nowhere to march.
Below the fort, from fields of silver water,
the moon rose on a chiton-fluted sea.

III

Perched on the low stern of the rented shallop,
he watched the barracks on the hill dilate
with every stroke behind the oarsman's ear.
The rower, silent, kept his gaze obliquely fixed on the wharf's receding beacon,
a mannequin with a skirt of lacy iron, and
in the opaque, slowly colouring harbour
the one sound was the plump plash of the oars,
each stroke concluding with the folded gurgle
of an intaken breath. Weakly protesting,
the oarlock's squeak, the gunwale's heaving lurch,
the pause upheld after each finished stroke,
unstudied, easy, pentametrical,
one action, and one thought. Halfway across
the chord between the downstroke of the oar
and its uplifted sigh was deepened
by a donkey's rusty winch, from Foux Lachaud,
a herring gull's one creak, till the bay grew
too heavy for reflection. The rower veered
precisely, triangulating his approach,
headed for an abandoned rocky inlet
that reeked of butchered turtles, then the shallop
skimmed shallow water, the coast sliding

past easily, easily sliding rocks and trees
over the mossed mosaic of bright stones,
making their arrival secret. He would remember
a child in a canted whaleboat rounding this harbour,
coves chopped in a crescent by the whale's jaw,
with "Boy" or "Babs" Monplaisir at the tiller,
now, lecherous, lecherous, sighed the insucked water,
muttered the wiry writhing sun-shot creeks
and grasses, lecherous, skittered the thin,
translucent minnows from the skiff's shadow,
you with your finger in the pie of sin,
you with an iron in the fire, tell her
that the house could speak. Odour of fish,
odour of lechery. Who spoke? I,
said the Indian woman you finger-poked in the doorway.

I, said the Negro whore on the drawing-room floor
under the silent portraits of your parents,
while Anna slept,
her golden body like a lamp blown out
that holds, just blown, the image of the flame.

    IV

Magical lagoon, stunned
by its own reflection!
The boat, the stone
pier, the water-odorous
boathouse, the trees
so quiet they were always
far. I could drown there,

as now, even now, an Ann,
Anna, Andreuille,
swims from a ring of whispering
young sisters, her head
emerging from their
swirling eddy.

The sixteen-year-old sun
plates her with light,
freckled, from now on
her colour only,

leaf-freckled forearm
brown of leafless April,
the russet hair, the freckled
big-boned wrist. Reader,

imagine the boat stayed,
the harbour stayed, the oar's
uplifted wand,

hold the light's changes to
a single light, repeat
the voyage, delay the arrival,
in that bright air,

he wished himself moving
yet forever there.
The disc of the world turned
slowly, she was its centre,

the chill of water entered

the shell of her palm,
in membranous twilight
the match of the first star

through the door of sunset always left ajar.
And all bread savoured
of her sunburnt nape,

her laughter a white napkin
shaken under the leaves.
We sit by the stone wall

all changes to grey stone,
stone hands, stone air,
stone eyes, from which

irisless, we stare,
wishing the sea were stone,
motion we could not hear.

No silence, since,
its equal.

For one late afternoon, when again she stood
in the door of a twilight always left ajar,
when dusk had softened the first bulb
the colour of the first weak star,
I asked her, "Choose,"
the amazed dusk held its breath,
the earth's pulse staggered,
she nodded, and that nod
married earth with lightning.

And now we were the first guests of the earth
and everything stood still for us to name.
Against the blades of palms and yellow sand,
I hear that open laugh,
I see her stride
as ruthless as that flax-bright harvester
Judith, with Holofernes' lantern in her hand.

V

And a vein opened in the earth,
its drops congealing into plum,
sorrel, and berry,
the year bleeding again, Noel, Noel,
blood for the bloodless birth,
blood deepening the poinsettia's Roman blades
after the Festival of the Innocents.
Life changing direction with the Trades.
A fresh wind, irrepressibly elate,
lifted the leaves' skirts, romped
down the Roman balconies,
polishing all that was already polished,
the sky, leaves, metal, and her face.
At nights in the Cantonment,
when the mouth of the full moon sang through the ruins,
choirs of black carollers patrolled the barracks
singing of holly and fresh-fallen snow,
among them, Anna,
profile of hammered gold,
head by Angelico,
stars choiring in gold leaf.

And Christmas came with its pretence at cold,
apples bubbled in barrows, the wind, beggarly
polished their cheeks;
along smoked kitchen walls, like a laboratory,
sweet, devilish concoctions, knuckles, roots,
blood-jellied jars, pungent and aromatic as the earth
which kept its secret till this season, were fermenting,
bark-knuckled ginger, the crimson bulbs
of sorrel like extinguished lights
packed up in last year's tinsel, sweated oil,
and the baked earth exuded
itself, as if, pebbled with clove,
it could at last be taken from its oven,
the night smelled like a cake
seasoned in anise, Falernum, and Madeira,
and the ruby glass of the chapels brimmed with wine.

VI

But this as well; some nights, after he left her,
his lechery like a mongrel nosed the ruins,
past Manoir's warehouse. In her absence
his nostrils prickled for the scent of sea-grapes;
Gregorias would laugh, "Drink, take a next sip.
You are creating this, and it will end.
The world is not like this,
nor is she, friend."

The cold glass was her lip.
Every room he entered was an album
from which her image had been crudely torn.

Chapter 14

—Anna awaking

I

When the oil green water glows but doesn't catch,
only its burnish, something wakes me early,
draws me out breezily to the pebbly shelf
of shallows where the water chuckles
and the ribbed boats sleep like children,
buoyed on their creases. I have nothing to do,
the burnished kettle is already polished,
to see my own blush burn,
and the last thing the breeze needs is my exhilaration.

I lie to my body with useless chores.
The ducks, if they ever slept, waddle knowingly.
The pleats of the shallows are neatly creased
and decorous and processional,
they arrive at our own harbour from the old hospital
across the harbour. When the first canoe,
silent, will not wave at me,
I understand, we are acknowledging
our separate silences, as the one silence,
I know that they know my peace as I know theirs.
I am amazed that the wind is tirelessly fresh.
The wind is older than the world.

It is always one thing at a time.
Now, it is always girlish.
I am happy enough to see it as a kind
of dimpled, impish smiling.
When the sleep-smelling house stirs
to that hoarse first cough, that child's first cry,
that rumbled, cavernous questioning of my mother,
I come out of the cave
like the wind emerging,
like a bride, to her first morning.

I shall make coffee.
The light, like a fiercer dawn,
will singe the downy edges of my hair,
and the heat will plate my forehead till it shines.
Its sweat will share the excitement of my cunning.
Mother, I am in love.
Harbour, I am waking.
I know the pain in your budding, nippled limes,
I know why your limbs shake, windless, pliant trees.
I shall grow grey as this light.
The first flush will pass.
But there will always be morning,
and I shall have this fever waken me,
whoever I lie to, lying close to, sleeping
like a ribbed boat in the last shallows of night.

But even if I love not him but the world,
and the wonder of the world in him, of him in the world,
and the wonder that he makes the world waken to me,
I shall never grow old in him,
I shall always be morning to him,

and I must walk and be gentle as morning.
Without knowing it, like the wind,
that cannot see her face,
the serene humility of her exultation,
that having straightened the silk sea smooth, having noticed
that the comical ducks ignore her, that
the childish pleats of the shallows are set straight,
that everyone, even the old, sleeps in innocence,
goes in nothing, naked, as I would be,
if I had her nakedness, her transparent body.
The bells garland my head. I could be happy,
just because today is Sunday. No, for more.

II

Then Sundays, smiling, carried in both hands
a towelled dish bubbling with the good life
whose fervour, steaming, beaded her clear brow,
from which damp skeins were brushed,
and ladled out her fullness to the brim.
And all those faded prints that pressed their scent
on her soft, house-warm body
glowed from her flesh with work,
her hands that held the burnish of dry hillsides
freckled with firelight,
hours that ripened till the fullest hour
could burst with peace.

"Let's go for a little walk," she said, one afternoon,
"I'm in a walking mood." Near the lagoon,
dark water's lens had made the trees one wood
arranged to frame this pair whose pace

unknowingly measured loss,
each face was set towards its character.
Where they now stood, others before had stood,
the same lens held them, the repeated wood,
then there grew on each one
the self-delighting, self-transfiguring stone
stare of the demi-god.
Stunned by their images they strolled on, content
that the black film of water kept the print
of their locked images when they passed on.

III

And which of them in time would be betrayed
was never questioned by that poetry
which breathed within the evening naturally,
but by the noble treachery of art
that looks for fear when it is least afraid,
that coldly takes the pulse-beat of the heart
in happiness; that praised its need to die
to the bright candour of the evening sky,
that preferred love to immortality;
so every step increased that subtlety
which hoped that their two bodies could be made
one body of immortal metaphor.
The hand she held already had betrayed
them by its longing for describing her.

Chapter 15

I

Still dreamt of, still missed,
especially on raw, rainy mornings, your face shifts
into anonymous schoolgirl faces, a punishment,
since sometimes you condescend to smile,
since at the corners of the smile there is forgiveness.

Besieged by sisters, you were a prize
of which they were too proud, circled
by the thorn thicket of their accusation,
what grave deep wrong, what wound have you brought, Anna?

The rain season comes with its load.
The half-year has travelled far. Its back hurts.
It drizzles wearily.

It is twenty years since,
after another war, the shell cases are where?
But in our brassy season, our imitation autumn,
your hair puts out its fire,
your gaze haunts innumerable photographs,

now clear, now indistinct,
all that pursuing generality,
that vengeful conspiracy with nature,

all that sly informing of objects,
and behind every line, your laugh
frozen into a lifeless photograph.

In that hair I could walk through the wheatfields of Russia,
your arms were downed and ripening pears,
for you became, in fact, another country,

you are Anna of the wheatfield and the weir,
you are Anna of the solid winter rain,
Anna of the smoky platform and the cold train,
in that war of absence, Anna of the steaming stations,

gone from the marsh edge,
from the drizzled shallows
puckering with gooseflesh,
Anna of the first green poems that startingly hardened,

of the mellowing breasts now,
Anna of the lurching, long flamingoes
of the harsh salt lingering in the thimble
of the bather's smile,

Anna of the darkened house, among the reeking shell cases,
lifting my hand and swearing us to her breast,
unbearably clear-eyed.

You are all Annas, enduring all goodbyes,
within the cynical station of your body,
Christie, Karenina, big-boned and passive,

that I found life within some novel's leaves

more real than you, already chosen
as his doomed heroine. You knew, you knew.

   II

Who were you, then?
The golden partisan of my young Revolution,
my braided, practical, seasoned commissar,

your back, bent at its tasks, in the blue kitchen,
or hanging flags of laundry, feeding the farm's chickens,
against a fantasy of birches,

poplars, or whatever.
As if a pen's eye could catch that virginal litheness,
as if shade and sunlight leoparding the blank page
could be so literal,

foreign as snow,
far away as first love,
my Akhmatova!

Twenty years later, in the odour of burnt shells,
you can remind me of "A Visit to the Pasternaks,"
so that you are suddenly the word "wheat,"

falling on the ear, against the frozen silence of a weir,
again you are bending
over a cabbage garden, tending
a snowdrift of rabbits,
or pulling down the clouds from the thrumming clotheslines.

If dreams are signs,
then something died this minute,
its breath blown from a different life,

from a dream of snow, from paper
to white paper flying, gulls and herons
following this plough. And now,

you are suddenly old, white-haired,
like the herons, the turned page. Anna, I wake
to the knowledge that things sunder
from themselves, like peeling bark,

to the emptiness
of a bright silence shining after thunder.

III

"Any island would drive you crazy,"
I knew you'd grow tired
of all that iconography of the sea

like the young wind, a bride
riffling daylong the ocean's catalogue
of shells and algae,

everything, this flock
of white, novitiate herons
I saw in the grass of a grey parish church,

like nurses, or young nuns after communion,

their sharp eyes sought me out
as yours once, only.

And you were heron-like,
a water-haunter,
you grew bored with your island,

till, finally, you took off,
without a cry,
a novice in your nurse's uniform,

years later I imagined you
walking through trees to some grey hospital,
serene communicant,
but never "lonely,"

like the wind, never to be married,
your faith like folded linen, a nun's, a nurse's,
why should you read this now?

No woman should read verses
twenty years late. You go about your calling, candle-like,
carrying yourself down a dark aisle

of wounded, married to the sick,
knowing one husband, pain,
only with the heron-flock, the rain,

the stone church, I remembered . . .
Besides, the slender, virginal New Year's
just married, like a birch
to a few crystal tears,

and like a birch bent at the register
who cannot, for a light's flash, change her name,
she still writes '65 for '66;

so, watching the tacit
ministering herons, each at its
work among the dead, the stone church, the stones,

I made this in your honour, when
vows and affections failing
your soul leapt like a heron sailing
from the salt, island grass

into another heaven.

    IV

Anna replies:

I am simple,
I was simpler then.
It was simplicity
which seemed so sensual.

What could I understand,
the world, the light? The light
in the mud-stained sea-wash,
the light in a gull's creak

letting the night in?
They were simple to me,
I was not within them as simply

as I was within you.

It was your selflessness
which loved me as the world,
I was a child, as much
as you, but you brought the tears

of too many contradictions,
I became a metaphor, but
believe me I was unsubtle as salt.

And I answer, Anna,
twenty years after,
a man lives half of life,
the second half is memory,

the first half, hesitation
for what should have happened
but could not, or

what happened with others
when it should not.

A gleam. Her burning grip. The brass shell cases,
oxidized, the brass reeking of cordite,
forty-one years after the Great War. The gleam
of brass reburnished in the allamanda,
through the barbed wire of bougainvillea thorns
beyond the window, on the sun-chevroned porch
I watched the far cannon smoke of cloud
above the Morne, wounded, struck dumb,
as she drew my hand firmly to the firstness

of the crisp, fragile cloth across her breast,
in a locked silence, she the nurse,
I the maimed soldier. There have been
other silences, none as deep. There has since
been possession, none as sure.

Chapter 16

—The cement phoenix

I

Meanwhile to one metre, in the burnt town
things found the memory of their former places,
that vase of roses slowly sought its centre
like a film reeled backward, like
a poltergeist reversed.
Oleographs of Christ the Sacred Heart
sailed towards their new hooks and anchored there,
doilies like feathers floating settled softly,
frames drew their portraits like a closing rose,
laces resumed their spinsterish precision
and parlours were once more varnished, sacrosanct.
Apartment blocks whitened the air,
cul-de-sacs changed their dialect patronyms
to boulevards and avenues,
the cement phoenix rose.
All day in the gutted roads of the new city
the cement mixers snarled American, white-faced
the city entered its half-century.

Slowly she rose, the New Jerusalem
created in the image of *The Commonwealth Today*,
a hearty brochure from Whitehall

showing Welfare Officers grinning behind prize bulls
at county fairs, the new Senior Civil Servant
folded in his greatcoat, the phoenix metaphor flew
from tongue to tongue.
                        New cement blocks
five or six stories high
in their didactic Welfare State severity,
boulevards short of breath
confronted the old town.
From the verandah of the old wooden college
I watched the turning pages of the sea.

I leant across the rakish balcony
smelling the sunlit iron of the burnt town.
Burnt flesh. Our blitz was over too. I felt
the voices of children under my feet.
"Saul has slain his thousand,
David his ten thousand."

The bones of our Hebraic faith were scattered
over such a desert, burnt and brackened gorse,
their war was over, it had not been
the formal tapestry bled white by decorum,
it had infected language,
gloria Dei and the glory of
the Jacobean Bible were the same. The shoes
of cherubs piled in pyramids
outside the Aryan ovens.

A ghost, accosting, softly grew beside me.
My other self, the Brother, the mathematical Poet,
scented with mint, he cited, softly, Perse:

"The beauty of this world hath made me sad."
I turned and studied him. Our smiles blent.
He possessed the bulging, tubercular stare,
the shadow-sooted eyes whose sockets held the hunger
of survivors of the death camps and the soap vats,
but, bony, muscular as his body was
and thinly sheeted with its film of sweat
it hinted of deeper emaciation,
a gnawing, lacerated elegance.
The Andrew Aguecheek hair thinned like flax
winnowed from the bumpy skull, a phrenologist's field day,
so wispy-frail, so oven-singed,
you feared the lightest breeze would sear him bald.
He, too, a poet once. His, too, this exile!
Never to leave his isle, till Mary called him,
to sit and watch the twilight in this harbour
igniting other lives, watching the herring gulls rehearse
with every dusk their cycles of departure,
to watch the visionary glare
tarnish to tin. To hear, waking at night,
the rain driving its nails into this ground,
into his hands, to walk the kelp-piled beach
and hear the waves arriving with stale news.
Go then, I thought, but the wave held back my hand,
the grass felt hurt, the stars shone without privilege,
in those soft dusks, like him,
it rained within me as it rained on Ireland.

While, in the dark hold below the coffin planks,
a generation of slaves' children sang,
"Where balmy breezes blow
Soft winds are playing . . .

Santaaa Lucheeeea
Santaaa Lucheeeea."
Steered now by Irish hands to their new epoch.

Other men's voices,
other men's lives and lines.

The accolade, the accolade.
Tea with the British Council Representative,
tannin, calfskin, gilt, and thank you vellum much,
of course you will soon shed your influences,
silvery cadence measured, the eavesdropping coarse vegetation
outside white jalousies, the indoor palms
nodding to Mr. Winters's approbation,
a rubicund, gurgling consul, "keener on music"
but capable of knowing talent when he sees it.
I am hoisted on silvery chords upward,
eager for the dropped names like sugar cubes.
Eliot. Plop. Benjamin Britten. Clunk. Elgar. Slurp.
Mrs. Winters's cheeks gleaming. Polished cherries.
Lawns. Elegance. Remembering elms. England, then. When?
Down on her speckled forearm. More tea.
Thank you, my mind burrowing her soft scented crotch.
First intimations of immortality.
Other men's wives.

II

Sister Annunziata
sat by the white wall
of the convent balcony
whose shadows alternated

like the piano keys
of her pupil's practice,
under its black snood
her starched brow, a wall
of ageing ivory,
listened as Anna played
in the rivering afternoon,
arpeggios of minnows
widening from her hand,
as through the chapel prism
Sister Gabriela
settled like a tired pigeon,
old nun whose steel glasses
flashed spears into her Christ,
Annunziata's head
a wall of virgin plaster
below which her dark eyes
burned fiercely in their niche.

And to the monody of piano practice
gables looped the skyline of the old town,
swifts scored the air,
jalousies shuttered like the singer's eyes,
but in those idling afternoons
the one melody more sorrowful
than some frail Irish air
was the old theme of smoke rising from back yards
barred with rusted tin, the thin
search for something that had surrendered soon.

I walked down to the wharf, as usual
looking seaward,

the island lowered
the sun, and rocked
slowly, at anchor.
The usual smoky twilight
blackened our galvanized roof with its nail holes of stars,
the old dark tarred the wharves,
the mesmerized dogfishes glared
through phosphorus and the stunned eels wriggled
in the ochre arclight of Prince Albert's Basin,
where the schooners smelt of islands,
and far out, the fireflies
or villas signalled an incoming liner
far as Christmas.
Earth-heart, I prayed,
nerves of raw fibre,
uproot me, yet
let what I have sworn to love not feel betrayed
when I must go, and, if I must go,
make of my heart an ark,
let my ribs bear
all, doubled by
memory, down to the emerald fly
marrying this hand, and be
the image of a young man on a pier,
his heart a ship within a
ship within a ship, a bottle
where this wharf, these
rotting roofs, this sea,
sail, sealed in glass.

III

To that vow were addressed all those vibrations of rain
like a railing echoing arpeggios where her fingers
sounded, as the day bending gathered
the fallen clouds before nightfall;
they were shaping their fallacy
in your breast, like an ancient engraving
of Italianate cabbage palms, where the feathery leaves
of flamboyants backed the old landscapes.
How often didn't you hesitate
between rose-flesh and sepia,
your blood like a serpent whispering
of a race incapable of subtler shadow,
of music, architecture, and a complex thought.
In that acid was evening etched,
a coppery glaze plated the landscape
till your envious anger accused it of being too poor;
but be glad that you were touched
by some other's sadness, that when your hand trembles
and the tightening railings sound,
or the sky, before rain, sounds like a monstrous shell
where their voices are, be happy
in every uncertainty. Cherish the stumbling
that lashes your eyes with branches,
that, threatened with rain,
your sorrow is still uncertain.

You had begun to write those letters to no one,
your friends, like leaves, seemed too preoccupied
with balance. You faced the blank page

and trembled, you had learnt by heart
the monotonous scrawl of the beaches
for years trying to reach you,
delivering the same message, Go,
in the crab's carapace from which the crabsoul had vanished,
in eyes ground the colour of sea-stone.
Their lives slipped into your own
like letters under a door.

Chapter 17

I

Note after note the year was orchestrating
those wires of manuscript ruled on its clouds
till they were black with swallows quavering
for their surge north. Lightning frequently
crackled across the watersheds, thunder
rattled the sky's tightened parchment.
He haunted beaches,
the horizon tightened round his throat.
Their wings flashing like tinfoil
the sanderlings refused his messages.
The islands were a string of barges towed nowhere,
every view
assembling itself to say farewell.

At Vieuxfort
the soldiers had broken up their base and gone,
the mustard-yellow bungalows through the palms
were empty or dismantled. Behind the tarred screens,
behind the rusting fly-wire, nothing stirred,
the runways cracked open like an idiot's smile.

The steady salt air
from the open Atlantic rusted bolts,
air hangars, latches, children's toys,

the wooden treaders gave, grey, plump with rot.
From Micoud to L'Anse Paradis
the breakers, like a louder silence, roared.

    II

There were ducks, supposedly,
that used the windward coast as a way station
on their haul south from as far as Sombreor
and La Isla de Pinos to Brazil.
I looked for them flying, stupidly, at daybreak,
I felt the instinct of their passage,
but I saw none.
              The only birds
were the pernickety, finical sanderlings,
their coats turned white for winter,
testing the shallows' edge,
their parallels as far as the wild ducks.
And still I saw
the possibility of angels,
suction of their bare heel razed from the sand,
in the wind-chased brightness after their shadows
raced from the tide, their flight followed so often
by the look-they-have-left-us impulse of sanderlings
over bright water, and herring gulls circled
the eddy where heaven had sucked them in,
I knew where they hid behind the walls of cloud,
I heard when they rushed from bent canes urgently
to bless some other island, I was the guest
of their impatient presence, one voice behind
my father's, one foot impressed in
the sand-indenting heel of a seraph,

I saw their colour,
in the steel, silver scales of the sea,
rasp-winged and hoarse in their fretful, marshalled flocks,
work-worn and restless, and when they turned
their gaze, it was not the glare of eagles,
they were not eternity's falcons,
but impatient, commissioned beings, their fierce
expressionless look benign and tolerant,
they were simple as gulls
not caring if they were noticed,
when we were gone, others
would watch them.
They were without revenge,
for those who left them
their only punishment was absence.

III

And so one summer after I returned, we arranged
to stay in the old village and we spent
two days and one night there, but except
for the first few hours it was somehow different,
as if either the island or myself had changed,
but not Gregorias, and we both spent
a bad night sleeping on our shoes for pillows,
hearing the rasping surf until the dawn,
but it was not the same at dawn, it was a book
you'd read a life ago walking up the brown
sand, the filth, and where the sea breaks at La Vierge,
a dead mind wandering at the long billows,
and I left there that morning with a last look
at things that would not say what they once meant.

IV

One dawn the sky was warm pink thinning to no colour.
In it, above the Morne, the last star shone,
measuring the island with its callipers.
As usual, everywhere, the sinuations of cockcrow,
a leisured, rusting, rising and falling,
echoed the mountain line. The day creaked
wearily open. A wash of meagre blue entered the sky.
The final star diminished and withdrew.
Day pivoted on a seagull's screeching hinge.
And the year closed. The allamandas fell,
medalling the shoulders of the last visitor.
At the airport, I looked towards the beach.
The sand had seen battalions come and go,
the vines had written their memorials,
all of that cannon fire taken up by cloud.
Nothing had altered the teal or mallard's route,
all that salt blood thinned out in the salt surf.
I shook Gregorias's hand. Dead almond leaf.
There was no history. No memory.
Rocks haunted by seabirds, that was all.
The house would survive, my brother would survive,
and yet how arrogant, how cruel
to think the island and Anna would survive
(since they were one), inviolate, under
their sacred and inverted bell of glass,
and that I was incapable of betrayal,
to imagine their lives revolving round my future,
to accept as natural their selfless surrender.
The three faces I had most dearly loved

that year, among the blurred faces in the crowd,
Gregorias laughing, "Jamaica just up the road, man,
just up the road." Harry hustling. Anna had not moved.
I watched the island narrowing, the fine
writing of foam around the precipices, then
the roads as small and casual as twine
thrown on its mountains, I watched till the plane
turned to the final north and turned above
the open channel with the grey sea between
the fishermen's islets until all that I love
folded in cloud. I watched the shallow green
breaking in places where there would be reef,
the silver glinting on the fuselage, each mile
tightening us and all fidelity strained
till space would snap it. Then, after a while
I thought of nothing; nothing, I prayed, would change.
When we set down at Seawell it had rained.

No metaphor, no metamorphosis,
as the charcoal-burner turns
into his door of smoke,
three lives dissolve in the imagination,
three loves, art, love, and death,
fade from a mirror clouding with this breath,
not one is real, they cannot live or die,
they all exist, they never have existed:

Harry, Dunstan, Andreuille.

Four
THE ESTRANGING SEA

> Who order'd that their longing's fire
> Should be, as soon as kindled, cool'd?
> Who renders vain their deep desire?—
> A God, a God their severance ruled!
> And bade betwixt their shores to be
> The unplumb'd, salt, estranging sea.
>
> ARNOLD
> "To Marguerite"

Chapter 18

—Noa Noa

I

You can't beat brushing young things from the country
in the country self. Might wind up there myself.
Black lissome limbs and teeth like fresh-cooked yams,
backs smooth and sleek like rainwashed aubergines
and tits like nippled naseberries, and he did that,
Gauguin had gone, Harry had built his mansion
upon the beached verge of the salt flood,
within a sea of roaring leaves, the gales
driving his houseboat deeper in the forest, buoyed
by green breakers of ocean the long night,
he had blacked out for their millennium,
a clock too tired to tell the time,
he had exchanged their future for the prime
simplicities of salt pork seasoning the pot,
for the white rum growling in his gut for lunch,
the smell of baked earth rising from the grid
of noon, for the cloud-cloth steaming in the tin
where the scale-backed breadfruit gurgled like a turtle
for another life, for jahbal and vahine.
Telling himself that although it stank
this was the vegetable excrement of natural life,
not their homogenized, chemical-ridden shit,
that there on the leafy ocean with his saints,

Vincent and Paul, his yellowing *Letters to Theo*
and *Noa Noa*, though the worms bored their gospel,
he no longer wanted what he could become,
his flame, made through their suffering, their flame,
nightly by the brass-haloed lamp, he prayed
whatever would come, come.
                        Why not, indeed? In deed.
There was his hand and the shadow of his hand,
there was his thought and the shadow of that thought
lying lighter than the shadow of a sound
across coarse canvas or the staring paper,
the quiet panic at the racing sun
his breath held before its trembling wick,
the done with its own horror of the undone
that frays us all to pieces and breakdown,
all of us, always, all ways, one after one.

He faced the canvas, bored
with the downseamed face of a man used
to giving orders, but before he began
the surface would acquire its old ambition.
It could not understand this newer life.
A spider began to thread
easel to bedstead.

           Now, where he had beheld
a community of graceful spirits
irradiating from his own control and centre,
through botany, history, lepidoptery, stamps,
his mind was cracking like the friable earth,
and in each chasm,
sprung nettles like the hands of certain friendships.

Sunset grew blacker in the fisherman's flesh.
He would resolve into a fish unless he left
the long beach darkening, for the village lamps,
with the claws of the furred sea gripping the high ground,
his eyes would phosphoresce, his head
bubble with legends through the fly-like heads
of fishnets slung between the poles,
where the palm trees were huge spiders stuck on shafts.
Fear rooted him. Run, like a child again, run, run!
The morning bleeds itself away,
everything he touches breaks,
like a child again, he reads
the legend of Midas and the golden touch,
from morning through the afternoon
he feels compelled to read
the enormous and fragile literature
of breakdown. It is like that visit
to that trembling girl, at whose quivering side,
her skin like a plagued foal's,
my own compassion quivered,
dark moons moving under her glazed eyelids,
who answers, "How was it?
It was all trembling."
It is fear and trembling.

II

Its initial intimation is
indifference to the uncontrollable,
to ease into the terrifying spaces
like Pascal's bloated body pricked out with stars,
to see, swimming towards us,

the enormous, lidless eyeball of the moon,
dumb, gibbering with silence, struck
by something it cannot answer
or the worst, the worst, an oceanic nothing,
it is all trembling,
it is fear and trembling.
It is the uncontrollable
persistence of the heartbeat,
the sweat, the twitching eye,
the hand from which the brush slips on its own,
the finical signals of an overdue defeat,
to tire of life, and yet not wish to die.

And yet, and yet,
the same day will persist in being good,
either its nature has not learnt to forget,
or fear was a faith it never understood;
it hoisted, stubbornly, its yards and spinnakers,
believed immortally in blue,
and had, as usual, the old engine, love,
nodded acknowledgement to the supreme maker's
hand, yet could not tell you why, or how it moved,
its bow of daylight driving to the dark,
as if its love
and the stunned blue afternoon were life enough.

III

Irascibility, muse of middle age.
How often you have felt you have wasted your life
among a people with no moral centre,
to want to move from the contagion of too many friends,

the heart congealing into stone,
how many would prefer to this poem
to see you drunken in a gutter,
and to catch in the corner of their workrooms
the uncertified odour of your death?
And perhaps, master, you saw early
what brotherhood means among the spawn of slaves
hassling for return trips on the middle passage,
spitting on their own poets,
preferring their painters drunkards,
for their solemn catalogue of suicides,
as I draw nearer your desolation, Cesar Vallejo,
and its raining Thursdays.

Do not tell me the world is the same,
that life is hard as a stone,
for I have known it when it was a flower
potent, annihilating with promise.
That the worst of us are wolves.
I no longer care for whom I write,
as you found in your hand, sir,
that terrible paralysis of their vindication
that out of such a man,
nothing would come,
they said that, and were already composing,
some by a sentence, some by a phrase, some by their spit,
but most with a dry remark, like a fistful of dirt
flung into your grave, from such a man
what would you expect,
but a couple of paintings
and a dog's life?

   I would refrain,
I withhold from myself that curse,
but in this battle, it is them, or me,
and as it was you who lost,
and they who pitied your losing,
and they who deny now their victory,
it is, it is sad, though, a struggle,
without engagement.

  IV

Let this one remember how
I closed with gentleness,
as if I were his brother,
with all the love that I had left
his sister's young eyes,
and the same one, the *chauve*-head
with the harelip and the lisp
that is, if I had only known, a serpent's,
certain matters of money,
certain matters of preferment,
and so many other embarrassments,
but I need to write them,
or I myself would not believe
that the world has left such men,
that the race is still a stigma,
that the truth is nothing more
than a puddle of clear water
dammed in a ditch. Still, master, I cannot
enter the inertia of silence.
My hands, like those of a madman's,

cannot be tied. I have no friends
but the oldest, words. This, at least,
master, none can take from me.
But the path increases with snakes.

No,
think of the weight which
the delicate blades of the fern endure,
the weight of the world, and
everything else in its world that is not
fern, yet it can be eased from
earth by a fist, rainstorms richen its
roots, what it takes from wind
is hard to believe, but its sweat
gleams, it is chained in its own dew,
it is locked into earth, unlike the delicate
ribs of some men. Uprooted
they quail.

For here, what was success?
It was the mean, inner
excitement at having survived.
Had he been freed? Or
had heart, guts, and talent
exhausted him? Every muscle
ached like a rusting hawser
to hoist him heavenward towards
his name, pierced with the stars
of Raphael, St. Greco, and later,
not stars, but the people's medals,
with Siqueiros, Gauguin, Orozco,
St. Vincent and St. Paul.
It would be worth it to fall,

with the meteor's orange brushstroke
from a falling hand, to hope
there is painting in heaven.

The Muse of inanition, the dead nerve,
where was the world in which we felt the centre,
our *mundo nuevo,* handed to black hands
copying the old laws, the new mistakes? Where?
Speak to the Indian grazing his two cows
across the allotment.

"The brown one now [the cow], she is in season,
she was calling last moon."
"Why you don't ask the government for a farm?"
"Well, I apply, but all dem big boys so, dem ministers,
dem have their side. Cockroach must step aside
to give fowl chance." Ah, brave third world!

Cockroach must step aside to give fowl chance.
Leaves of the long afternoon silverly trembling.

Chapter 19

—Frescoes of the New World II

I

My Anna, my Beatrice,
I enclose in this circle of hell,
in the stench of their own sulphur of self-hatred,
in the steaming, scabrous rocks of Soufrière,
in the boiling, pustular volcanoes of the South,
all o' dem big boys, so, dem ministers,
ministers of culture, ministers of development,
the green blacks, and their old toms,
and all the syntactical apologists of the Third World
explaining why their artists die
by their own hands, magicians of the New Vision.
Screaming the same shit.
Those who peel, from their own leprous flesh, their names,
who chafe and nurture the scars of rusted chains,
like primates favouring scabs, those who charge tickets
for another free ride on the middle passage,
those who explain to the peasant why he is African,
their catamites and eunuchs banging tambourines,
whores with slave bangles banging tambourines,
and the academics crouched like rats
listening to tambourines
jackals and rodents feathering their holes

hoarding the sea-glass of their ancestors' eyes,
sea-lice, sea-parasites on the ancestral sea-wrack,
whose god is history. *Pax.*
Who want a new art,
and their artists dying in the old way.
Those whose promises drip from their mouths like pus.
Geryons gnawing their own children.
These are the dividers,
they encompass our history,
in their hands is the body
of my friend and the future,
they measure the skulls with callipers
and pronounce their measure
of toms, of traitors, of traditionals and Afro-Saxons.
They measure them carefully
as others once measured the teeth
of men and horses, they measure and divide.
Their music comes from the rattle of coral bones,
their eyes like worms drill into parchments,
they measure each other's sores
to boast who has suffered most,
and their artists keep dying,
they are the saints of self-torture,
their stars are pimples of pus
on the night of our grandfathers,
they are hired like dogs to lick the sores of their people,
their vision blurs, their future is clouded with cataract,
but out of its mist, one man,
whom they will not recognize, emerges
and staggers towards his lineaments.

Chapter 20

> —*Down their carved names*
> *the raindrop ploughs*
> —HARDY

I

Smug, behind glass, we watch the passengers,
like cattle breaking, disembark.
One life, one marriage later I watched Gregorias stride
across the tarmac at Piarco, that familiar lope
that melancholy hunter's stride
seemed broken, part of the herd.
                      Something inside
me broke subtly, like a vein. I saw him grope
desperately, vaguely for his friend,
for something which a life's bewilderment could claim
as stable. I shouted, "Apilo!"
Panic and wonder struggled for the grin.

"O the years, O . . ."
                      The highway canes unrolled in
silence past the car glass, like glass
the years divided. We fished for the right level, shrill,
hysterical, until, when it subsided,
a cautionary silence glazed each word.
Was he as broken down as I had heard,

driven deep in debt,
unable to hold down a job, painting so badly
that those who swore his genius vindicated
everything once, now saw it as a promise never kept?
Viciously, near tears, I wished him dead.

I wished him a spiteful martyrdom, in revenge
for their contempt, their tiring laughter.
After I told him, he laughed and said, "I tried it once.

"One morning I lay helplessly in bed,
everything drained, gone. The children crying.
I couldn't take any more. I had dreamed of dying.
I sent for Peggy, you remember her?
She's in the States now. Anyhow,
I sent her to the bathroom for a blade . . .
When she had brought it, I asked her to go.
I lay there with the razor blade in my hand . . .
I tried to cut my wrist . . . I don't know why
I stopped. I wanted very, very much to die . . .
Only some nights before, I had had a dream . . .
I dreamt . . ."
           And what use what he dreamt?
"We lived in a society which denied itself heroes"
(Naipaul), poor scarred carapace
shining from those abrasions it has weathered,
wearing his own humility like a climate,
a man exhausted, racked by his own strength,
Gregorias, I saw, had entered life.

They shine, they shine,
such men. After the vision

of their own self-exhaustion bores them,
till, slowly unsurprised at their own greatness,
needing neither martyrdom nor magnificence,
"I see, I see," is what Gregorias cried,
living within that moment where he died.

Re-reading Pasternak's *Safe Conduct*
as always again when life
startles under the lamplight,
I saw him brutally as Mayakovsky,
nostalgia, contempt raged for his death,
and the old choir of frogs,
those spinsterish, crackling cicadas.
Yet, even in such books
the element has burnt out,
honour and revelation are
a votive flame, and what's left
is too much like a wreath,
a smoky, abrupt recollection.
I write of a man whom life,
not death or memory, grants fame,
in my own pantheon, so, while
this fiery particle
thrives fiercely in another,
even if fuelled by liquour
to venerate the good,
honour the humbly great,
to render in "an irresponsible citizen"
the simple flame.

Too late, too late.

II

The rain falls like knives
on the kitchen floor.
The sky's heavy drawer
was pulled out too suddenly.
The raw season is on us.

For days it has huddled on the kitchen sill,
tense, a smoke-and-orange kitten
flexing its haunches,
coiling its yellow scream,
and now, it springs.
Nimble fingers of lightning
have picked the watershed,
the wires fling their beads.
Tears, like slow crystal beetles, crawl the pane.

On such days, when the postman's bicycle
whirrs drily like the locust
that brings rain, I dread my premonitions.
A grey spot, a waterdrop
blisters my hand.
A sodden letter thunders in my hand.
The insect gnaws steadily at its leaf,
an eaten letter crumbles in my hand,
as he once held my drawing to his face,
as though dusk were myopic, not his gaze.

"Harry has killed himself. He was found dead
in a house in the country. He was dead for two days."

III

The fishermen, like thieves, shake out their silver,
the lithe knives wriggle on the drying sand.
They go about their work,
their chronicler has gone about his work.

At Garand, at Piaille, at L'Anse la Verdure,
the sky is grey as pewter, without meaning.
It thunders and the kitten scuttles back
into the kitchen bin
of coal, its tines sheathing, unsheathing,
its yellow eyes the colour of fool's gold.

He had left this note.
No meaning, and no meaning.

All day, on the tin roofs
the rain berates the poverty of life,
all day the sunset bleeds like a cut wrist.

IV

Well, there you have your seasons, prodigy!
For instance, the autumnal fall of bodies,
deaths, like a comic, brutal repetition,
and in the Book of Hours, that seemed so far,
the light and amber of another life,
there is a Reaper busy about his wheat,
one who stalks nearer, and will not look up
from the scythe's swish in the orange evening grass,

and the fly at the font of your ear
sings, Hurry, hurry!
Never to set eyes on this page,
ah Harry, never to read our names,
like a stone blurred with tears I could not read
among the pilgrims, and the mooning child
staring from the window of the high studio.

Brown, balding, with a lacertilian
jut to his underlip,
with spectacles thick as a glass paperweight
and squat, blunt fingers,
waspish, austere, swift with asperities,
with a dimpled pot for a belly from the red clay of Piaille.
Eyes like the glint of sea-smoothed bottle glass,
his knee-high khaki stockings,
brown shoes lacquered even in desolation.

People entered his understanding
like a wayside country church,
they had built him themselves.
It was they who had smoothed the wall
of his clay-coloured forehead,
who made of his rotundity an earthy
useful object
holding the clear water of their simple troubles,
he who returned their tribal names
to the adze, mattock, midden, and cooking pot.

A tang of white rum on the tongue of the mandolin,
a young bay, parting its mouth,
a heron silently named or a night-moth,
or the names of villages plaited into one map,

in the evocation of scrubbed back-yard smoke,
and he is a man no more
but the fervour and intelligence
of a whole country.

Leonce, Placide, Alcindor,
Dominic, from whose plane vowels were shorn
odorous as forest,
ask the charcoal-burner to look up
with his singed eyes,
ask the lip-cracked fisherman three miles at sea
with nothing between him and Dahomey's coast
to dip rainwater over his parched boards
for Monsieur Simmons, *pour* Msieu Harry Simmons,
let the husker on his pyramid of coconuts
rest on his tree.

Blow out the eyes in the unfinished portraits.

And the old woman who danced
with a spine like the "glory cedar,"
so lissome that her veins bulged evenly
upon the tightened drumskin of the earth,
her feet nimbler than the drummer's fingers,
let her sit in her corner and become evening
for a man the colour of her earth,
for a cracked claypot full of idle brushes,
and the tubes curl and harden,
except the red,
except the virulent red!

His island forest, open and enclose him
like a rare butterfly between its leaves.

Chapter 21

I

Why?
You want to know why?
Go down to the shacks then,
like shattered staves
bound in old wire
at the hour when
the sun's wrist bleeds in
the basin of the sea,
and you will sense it,

or follow the path
of the caked piglet through
the sea-village's midden,
past the repeated
detonations of spray,
where the death rattle
gargles in the shale,
and the crab,
like a letter, slides
into its crevice,
and you may understand this,

smell the late, ineradicable reek
of stale rags like rivers
at daybreak, or the dark corner

of the salt-caked shop where the cod
barrel smells of old women,
and you can start then,

to know how the vise
of horizon tightens
the throat, when the first sulphur star
catches the hum
of insects round the gas lantern
like flies round a sore.
No more? Then hang round the lobby
of the one cinema too early

in the hour between two illusions
where you startle at the chuckle
of water under the shallop
of the old schooner basin,
or else it is still under all
the frighteningly formal
marches of banana groves,
the smell from the armpits of cocoa,

from the dead, open mouths
of husked nuts
on the long beach at twilight,
old mouths filled with water,
or else with no more to say.

    II

So you have ceased to ask yourself,
nor do these things ask you,
for the bush too is an answer

without a question,
as the sea is a question, chafing,
impatient for answers,
and we are the same.
They do not ask us, master,
do you accept this?
A nature reduced to the service
of praising or humbling men,
there is a yes without a question,
there is assent founded on ignorance,
in the mangroves plunged to the wrist, repeating
the mangroves plunging to the wrist,
there are spaces
wider than conscience.

Yet, when I continue to see
the young deaths of others,
even of lean old men, perpetually young,
when the alphabet I learnt as a child
will not keep its order,
see the young wife, self-slain
like scentful clove in the earth,
a skin the colour of cinnamon,
there is something which balances,
I see him bent under the weight of the morning,
against its shafts,
devout, angelical,
the easel rifling his shoulder,
the master of Gregorias and myself,
I see him standing over the bleached roofs
of the salt-streaked villages,
each steeple pricked
by its own wooden star.

I who dressed too early for the funeral of this life,
who saw them all, as pilgrims of the night.

III

And do I still love her, as I love you?
I have loved all women who have evolved from her,
fired by two marriages
to have her gold ring true.
And on that hill, that evening,
when the deep valley grew blue with forgetting,
why did I weep,
why did I kneel,
whom did I thank?
I knelt because I was my mother,
I was the well of the world,
I wore the stars on my skin,
I endured no reflections,
my sign was water,
tears, and the sea,
my sign was Janus,
I saw with twin heads,
and everything I say is contradicted.

I was fluent as water,
I would escape
with the linear elation of an eel,
a vase of water in its vase of clay,
my clear tongue licked the freshness of the earth,
and when I leapt from that shelf
of rock, an abounding bolt of lace,
I leapt for the pride of that race
at Sauteurs! An urge more than mine,

so, see them as heroes or as the Gadarene swine,
let it be written, I shared, I shared,
I was struck like rock, and I opened
to His gift!
I laughed at my death-gasp in the rattle
of the sea-shoal.
You want to see my medals? Ask the stars.
You want to hear my history? Ask the sea.
And you, master and friend,
forgive me!
Forgive me, if this sketch should ever thrive,
or profit from your gentle, generous spirit.
When I began this work, you were alive,
and with one stroke, you have completed it!

O simultaneous stroke of chord and light,
O tightened nerves to which the soul vibrates,
some flash of lime-green water, edged with white—
"I have swallowed all my hates."

IV

For I have married one whose darkness is a tree,
bayed in whose arms I bring my stifled howl,
love and forgive me!
Who holds my fears at dusk like birds which take
the lost or moonlit colour of her leaves,
in whom our children
and the children of friends settle
simply, like rhymes,
in whose side, in the grim times
when I cannot see light for the deep leaves,
sharing her depth, the whole lee ocean grieves.

Chapter 22

I

Miasma, acedia, the enervations of damp,
as the teeth of the mould gnaw, greening the carious stump
of the beaten, corrugated silver of the marsh light,
where the red heron hides, without a secret,
as the cordage of mangrove tightens
bland water to bland sky
heavy and sodden as canvas,
where the pirogue foundered with its caved-in stomach
(a hulk, trying hard to look like
a paleolithic, half-gnawed memory of pre-history)
as the too green acid grasses set the salt teeth on edge,
acids and russets and water-coloured water,
let the historian go mad there
from thirst. Slowly the water rat takes up its reed pen
and scribbles. Leisurely, the egret
on the mud tablet stamps its hieroglyph.

The explorer stumbles out of the bush crying out for myth.
The tired slave vomits his past.
The Mediterranean accountant, with the nose of the water rat,
ideograph of the egret's foot,
calculates his tables,
his eyes reddening like evening in the glare of the brass lamp;
the Chinese grocer's smile is leaden with boredom:
so many lbs. of cod,

              so many bales of biscuits,
on spiked shop paper,
the mummified odour of onions,
spikenard, and old Pharaohs peeling like onionskin
to the archaeologist's finger—all that
is the Muse of history. Potsherds,
and the crusted amphora of cutthroats.

Like old leather,
tannic, stinking, peeling in a self-contemptuous
curl away from itself,
the yellowing poems, the spiked brown paper,
the myth of the golden Carib,
like a worn-out film,
the lyrical arrow in the writhing Arawak maiden
broken under the leaf-light.
                              The astigmatic geologist
stoops, with the crouch of the heron,
deciphering—not a sign.
All of the epics are blown away with the leaves,
blown with the careful calculations on brown paper;
these were the only epics: the leaves.

No horsemen here, no cuirasses
crashing, no fork-bearded Castilians,
only the narrow, silvery creeks of sadness
like the snail's trail,
only the historian deciphering, in invisible ink,
its patient slime,
no cataracts abounding down gorges
like bolts of lace,
while the lizards are taking a million years to change,
and the lopped head of the coconut rolls to gasp on the sand,

its mouth open at the very moment
of forgetting its name.

That child who sets his half-shell afloat
in the brown creek that is Rampanalgas River—
my son first, then two daughters—
towards the roar of waters,
towards the Atlantic with a dead almond leaf for a sail,
with a twig for a mast,
was, like his father, this child,
a child without history, without knowledge of its pre-world,
only the knowledge of water runnelling rocks,
and the desperate whelk that grips the rock's outcrop
like a man whom the waves can never wash overboard;
that child who puts the shell's howl to his ear,
hears nothing, hears everything
that the historian cannot hear, the howls
of all the races that crossed the water,
the howls of grandfathers drowned
in that intricately swivelled Babel,
hears the fellaheen, the Madrasi, the Mandingo, the Ashanti,
yes, and hears also the echoing green fissures of Canton,
and thousands without longing for this other shore
by the mud tablets of the Indian provinces,
robed ghostly white and brown, the twigs of uplifted hands,
of manacles, mantras, of a thousand kaddishes,
whorled, drilling into the shell,
see, in the evening light by the saffron, sacred Benares,
how they are lifting like herons,
robed ghostly white and brown,
and the crossing of water has erased their memories.
And the sea, which is always the same,
accepts them.

And the shore, which is always the same,
accepts them.

In the shallop of the shell,
in the round prayer,
in the palate of the conch,
in the dead sail of the almond leaf
are all of the voyages.

    II

And those who gild cruelty,
who read from the entrails of disembowelled Aztecs
the colors of Hispanic glory
greater than Greece,
greater than Rome,
than the purple of Christ's blood,
the golden excrement on barbarous altars
of their beaked and feathered king,
and the feasts of human flesh,
those who remain fascinated,
in attitudes of prayer,
by the festering roses made from their fathers' manacles,
or upraise their silver chalices flecked with vomit,
who see a golden, cruel, hawk-bright glory
in the conquistador's malarial eye,
crying, at least here
something happened—
they will absolve us, perhaps, if we begin again,
from what we have always known, nothing,
from that carnal slime of the garden,
from the incarnate subtlety of the snake,
from the Egyptian moment of the heron's foot

on the mud's entablature,
by this augury of ibises
flying at evening from the melting trees,
while the silver-hammered charger of the marsh light
brings towards us, again and again, in beaten scrolls,
nothing, then nothing,
and then nothing.

III

                Here, rest. Rest, heaven. Rest, hell.
Patchwork, sunfloor, seafloor of pebbles at Resthaven, Rampanalgas.
Sick of black angst.
Too many penitential histories passing
for poems. Avoid:
                1857 Lucknow and Cawnpore.
The process of history machined through fact,
for the poet's cheap alcohol,
lines like the sugarcane factory's mechanization of myth
ground into rubbish.
                1834 slavery abolished.
A century later slavishly revived
for the nose of the water rat, for the literature of the factory,
in the masochistic veneration of
chains, and the broken rum jugs of cutthroats.
Exegesis, exegesis, writers
giving their own sons homework.

Ratoon, ratoon,
immigrant hordes downed soughing,
sickled by fever, *mal d'estomac*,
earth-eating slaves fitted with masks against despair,
not mental despondence but helminthiasis.

Pour la dernière fois, nommez! Nommez!

Abouberika Torre commonly called Joseph Samson.
Hammadi Torrouke commonly called Louis Modeste.
Mandingo sergeants offered Africa back,
the boring process of repatriation,
while to the indentured Indians
the plains of Caroni seemed like the Gangetic plain,
our fathers' bones. Which father?

Burned in the pyre of the sun.
On the ashpit of the sand.
Also you, Grandfather. Rest, heaven, rest, hell.
I sit in the roar of that sun
like a lotus yogi folded on his bed of coals,
my head is circled with a ring of fire.

    IV

O sun, on that morning,
did I not mutter towards your
holy, repetitive resurrection, "Hare,
hare Krishna," and then, politely,
"Thank you, life"? Not
to enter the knowledge of God
but to know that His name
had lain too familiar on my tongue,
as this one would say "bread,"
or "sun," or "wine," I staggered,
shaken at my remorse, as one
would say "bride," or "bread,"
or "sun," or "wine," to believe—
and that you would rise again,

when I am not here, to catch
the air afire, that you need not
look for me, or need this prayer.

    V

So, I shall repeat myself,
prayer, same prayer, towards fire, same fire,
as the sun repeats itself and the thundering waters

for what else is there
but books, books and the sea,
verandahs and the pages of the sea,
to write of the wind and the memory of wind-whipped hair
in the sun, the colour of fire?

I was eighteen then, now I am forty-one,
I have had a serpent for companion,
I was a heart full of knives,
but, my son, my sun,

holy is Rampanalgas and its high-circling hawks,
holy are the rusted, tortured, rust-caked, blind almond trees,
your great-grandfather's, and your father's torturing limbs,
holy the small, almond-leaf-shadowed bridge
by the small red shop, where everything smells of salt,
and holiest the break of the blue sea below the trees,
and the rock that takes blows on its back
and is more rock,
and the tireless hoarse anger of the waters
by which I can walk calm, a renewed, exhausted man,
balanced at its edge by the weight of two dear daughters.

VI

Holy were you, Margaret,
and holy our calm.
What can I do now

but sit in the sun to burn
with an ageing mirror that blinds,
combing, uncombing my hair—

escape? No, I am inured
only to the real, which
burns. Like the flesh

of my children afire.
Inured. Inward. As rock,
I wish, as the real

rock I make real,
to have burnt out desire,
lust, except for the sun

with her corona of fire.
Anna, I wanted to grow white-haired
as the wave, with a wrinkled

brown rock's face, salted,
seamed, an old poet,
facing the wind

and nothing, which is,
the loud world in his mind.

Chapter 23

I

At the Malabar Hotel cottage
I would wake every morning surprised
by the framed yellow jungle of
the groyned mangroves meeting
the groyned mangroves repeating
their unbroken water line.
Years. The island had not moved
from anchor.
                  Generations of waves,
generations of grass, like foam
petalled and perished in an instant.

I lolled in the shallows like an ageing hammerhead
afraid of my own shadow, hungering there.
When my foot struck sand, the sky rang,
as I inhaled, a million leaves drew inward.
I bent towards what I remembered,
all was inevitably shrunken,
it was I who first extended my hand
to nameless arthritic twigs,
and a bush would turn in the wind
with a toothless giggle, and
certain roots refused English.
But I was the one in awe.

This was a new pain,
I mean the mimosa's averring
"You mightn't remember me,"
like the scars of that scrofulous sea-grape
where Gregorias had crucified a canvas,
and there, still dancing like the old woman,
was the glory, the *gloricidia*.
I would not call up Anna.
I would not visit his grave.

II

They had not changed, they knew only
the autumnal hint of hotel rooms,
the sea's engine of air-conditioners,
and the waitress in national costume
and the horsemen galloping past the single wave
across the line of Martinique, the horse or *la mer*
out of Gauguin by the Tourist Board.
Hotel, hotel, hotel, hotel, hotel and a club: The Bitter End.
This is not bitter, it is harder
to be a prodigal than a stranger.

III

I looked from old verandahs at
verandahs, sails, the eternal summer sea
like a book left open by an absent master.
And what if it's all gone,
the hill's cut away for more tarmac,
the groves all sawn,
and bungalows proliferate on the scarred, hacked hillside,

the magical lagoon drained
for the Higher Purchase plan,
and they've bulldozed and bowdlerized our Vigie,
our *ocelle insularum,* our Sirmio
for a pink and pastel NewTown where the shacks and huts stood
teetering and tough in unabashed unhope,
as twilight like amnesia blues the slope,
when over the untroubled ocean, the moon
will always swing its lantern
and evening fold the pages of the sea,
and peer like my lost reader silently
between the turning leaves
for the lost names
of Caribs, slaves, and fishermen?

Forgive me, you folk,
who exercise a patience
subtler, stronger than the muscles
in the wave's wrist,
and you, sea, with the mouth
of that old gravekeeper
white-headed, lantern-jawed,
forgive our desertions, you islands
whose names dissolve like sugar
in a child's mouth. And you, Gregorias.
And you, Anna. Rest.

   IV

But, ah Gregorias,
I christened you with that Greek name because
it echoes the blest thunders of the surf,

because you painted our first, primitive frescoes,
because it sounds explosive,
a black Greek's! A sun that stands back
from the fire of itself, not shamed, prizing
its shadow, watching it blaze!
You sometimes dance with that destructive frenzy
that made our years one fire.
Gregorias, listen, lit,
we were the light of the world!
We were blest with a virginal, unpainted world
with Adam's task of giving things their names,
with the smooth white walls of clouds and villages
where you devised your inexhaustible,
impossible Renaissance,
brown cherubs of Giotto and Masaccio,
with the salt wind coming through the window,
smelling of turpentine, with nothing so old
that it could not be invented,
and set above it your crude wooden star,
its light compounded in that mortal glow:
Gregorias, Apilo!

April 1965–April 1972

From

SEA GRAPES

[1976]

Sea Grapes

That sail which leans on light,
tired of islands,
a schooner beating up the Caribbean

for home, could be Odysseus,
home-bound on the Aegean;
that father and husband's

longing, under gnarled sour grapes, is
like the adulterer hearing Nausicaa's name
in every gull's outcry.

This brings nobody peace. The ancient war
between obsession and responsibility
will never finish and has been the same

for the sea-wanderer or the one on shore
now wriggling on his sandals to walk home,
since Troy sighed its last flame,

and the blind giant's boulder heaved the trough
from whose groundswell the great hexameters come
to the conclusions of exhausted surf.

The classics can console. But not enough.

Sunday Lemons

Desolate lemons, hold
tight, in your bowl of earth,
the light to your bitter flesh,

let a lemon glare
be all your armour
this naked Sunday,

your inflexible light
bounce off the shields of apples
so real they seem waxen,

share your acid silence
with this woman's remembering
Sundays of other fruit,

till by concentration
you grow, a phalanx of helmets
braced for anything,

hexagonal cities where bees
died purely for sweetness,
your lamps be the last to go

on this polished table
this Sunday, which demands
more than the faith of candles

than helmeted conquistadors
dying like bees, multiplying
memories in her golden head;

as the afternoon vagues
into indigo, let your lamps
hold in this darkening earth

bowl, still life, but a life
beyond tears or the gaieties
of dew, the gay, neon damp

of the evening that blurs
the form of this woman lying,
a lemon, a flameless lamp.

New World

Then after Eden,
was there one surprise?
O yes, the awe of Adam
at the first bead of sweat.

Thenceforth, all flesh
had to be sown with salt,
to feel the edge of seasons,
fear and harvest,
joy that was difficult,
but was, at least, his own.

The snake? It would not rust
on its forked tree.
The snake admired labour,
it would not leave him alone.

And both would watch the leaves
silver the alder,
oaks yellowing October,
everything turning money.

So when Adam was exiled
to our New Eden, in the ark's gut,

the coined snake coiled there for good
fellowship also; that was willed.

Adam had an idea.
He and the snake would share
the loss of Eden for a profit.
So both made the New World. And it looked good.

Adam's Song

The adulteress stoned to death
is killed in our own time
by whispers, by the breath
that films her flesh with slime.

The first was Eve,
who horned God for the serpent,
for Adam's sake—which makes
everyone guilty or Eve innocent.

Nothing has changed,
for men still sing the song that Adam sang
against the world he lost to vipers,

the song to Eve
against his own damnation;
he sang it in the evening of the world

with the lights coming on in the eyes
of panthers in the peaceable kingdom
and his death coming out of the trees,

he sings it, frightened
of the jealousy of God and at the price
of his own death.

The song ascends to God, who wipes his eyes:

"Heart, you are in my heart as the bird rises,
heart, you are in my heart while the sun sleeps,
heart, you lie still in me as the dew is,
you weep within me, as the rain weeps."

Preparing for Exile

Why do I imagine the death of Mandelstam
among the yellowing coconuts,
why does my gift already look over its shoulder
for a shadow to fill the door
and pass this very page into eclipse?
Why does the moon increase into an arc-lamp
and the inkstain on my hand prepare to press thumb-downward
before a shrugging sergeant?
What is this new odour in the air
that was once salt, that smelt like lime at daybreak,
and my cat, I know I imagine it, leap from my path,
and my children's eyes already seem like horizons,
and all my poems, even this one, wish to hide?

Names

[*for Edward Brathwaite*]

I

My race began as the sea began,
with no nouns, and with no horizon,
with pebbles under my tongue,
with a different fix on the stars.

But now my race is here,
in the sad oil of Levantine eyes,
in the flags of the Indian fields.

I began with no memory,
I began with no future,
but I looked for that moment
when the mind was halved by a horizon.

I have never found that moment
when the mind was halved by a horizon—
for the goldsmith from Benares,
the stonecutter from Canton,
as a fishline sinks, the horizon
sinks in the memory.

Have we melted into a mirror,
leaving our souls behind?
The goldsmith from Benares,
the stonecutter from Canton,
the bronzesmith from Benin.

A sea-eagle screams from the rock,
and my race began like the osprey
with that cry,
that terrible vowel,
that I!

Behind us all the sky folded,
as history folds over a fishline,
and the foam foreclosed
with nothing in our hands

but this stick
to trace our names on the sand
which the sea erased again, to our indifference.

  II

And when they named these bays
bays,
was it nostalgia or irony?

In the uncombed forest,
in uncultivated grass
where was there elegance
except in their mockery?

Where were the courts of Castille?
Versailles' colonnades
supplanted by cabbage palms
with Corinthian crests,
belittling diminutives,
then, little Versailles
meant plans for a pigsty,
names for the sour apples
and green grapes
of their exile.

Their memory turned acid
but the names held;
Valencia glows
with the lanterns of oranges,
Mayaro's
charred candelabra of cocoa.
Being men, they could not live
except they first presumed
the right of every thing to be a noun.
The African acquiesced,
repeated, and changed them.

Listen, my children, say:
moubain: the hogplum,
cerise: the wild cherry,
baie-la: the bay,
with the fresh green voices
they were once themselves
in the way the wind bends
our natural inflections.

These palms are greater than Versailles,
for no man made them,
their fallen columns greater than Castille,
no man unmade them
except the worm, who has no helmet,
but was always the emperor,

and children, look at these stars
over Valencia's forest!

Not Orion,
not Betelgeuse,
tell me, what do they look like?
Answer, you damned little Arabs!
Sir, fireflies caught in molasses.

Sainte Lucie

I

The Villages

Laborie, Choiseul, Vieuxfort, Dennery,
from these sun-bleached villages
where the church bell caves in the sides
of one grey-scurfed shack that is shuttered
with warped boards, with rust,
with crabs crawling under the house-shadow
where the children played house;
a net rotting among cans, the sea-net
of sunlight trolling the shallows
catching nothing all afternoon,
from these I am growing no nearer
to what secret eluded the children
under the house-shade, in the far bell, the noon's
stunned amethystine sea,
something always being missed
between the floating shadow and the pelican
in the smoke from over the next bay
in that shack on the lip of the sandspit
whatever the seagulls cried out for
through the grey drifting ladders of rain
and the great grey tree of the waterspout,
for which the dolphins kept diving, that
should have rounded the day.

II

Pomme arac,
otaheite apple,
pomme cythère,
pomme granate,
moubain,
z'anananas
the pineapple's
Aztec helmet,
pomme,
I have forgotten
what pomme for
the Irish potato,
cerise,
the cherry,
z'aman
sea-almonds
by the crisp
sea-bursts,
au bord de la 'ouvière.
Come back to me,
my language.
Come back,
cacao,
grigri,
solitaire,
ciseau
the scissor-bird
no nightingales
except, once,
in the indigo mountains

of Jamaica, blue depth,
deep as coffee,
flicker of pimento,
the shaft light
on a yellow ackee
the bark alone bare
jardins
en montagnes
en haut betassion
the wet leather reek
of the hill donkey.

Evening opens at
a text of fireflies,
in the mountain huts
ti cailles betassion
candles,
candleflies
the black night bending
cups in its hard palms
cool thin water
this is important water,
important?
imported?
water is important
also very important
the red rust drum
the evening deep
as coffee
the morning powerful
important coffee

the villages shut
all day in the sun.

In the empty schoolyard
teacher dead today
the fruit rotting
yellow on the ground,
dyes from Gauguin
the pomme arac dyes
the earth purple,
the ochre roads
still waiting in the sun
for my shadow,
Oh, so you is Walcott?
you is Roddy brother?
Teacher Alix son?
and the small rivers
with important names.

And the important corporal
in the country station
en betassion
looking towards the thick
green slopes of cocoa
the sun that melts
the asphalt at noon,
and the woman in the shade
of the breadfruit bent over
the lip of the valley,
below her, blue-green
the lost, lost valleys

of sugar, the bus rides,
the fields of bananas
the tanker still rusts
in the lagoon at Roseau,
and around what corner
was uttered a single
yellow leaf,
from the frangipani
a tough bark, reticent,
but when it flowers
delivers hard lilies,
pungent, recalling

Martina, or Eunice
or Lucilla,
who comes down the steps
with the cool, side flow
as spring water eases
over shelves of rock
in some green ferny hole
by the road in the mountains,
her smile like the whole country,
her smell, earth,
red-brown earth, her armpits
a reaping, her arms
saplings, an old woman
that she is now,
with other generations
of daughters flowing
down the steps,
gens betassion,
belle ti fille betassion,

until their teeth go,
and all the rest.

O Martinas, Lucillas,
I'm a wild golden apple
that will burst with love
of you and your men,
those I never told enough
with my young poet's eyes
crazy with the country,
generations going,
generations gone,
moi c'est gens Ste. Lucie.
C'est la moi sorti;
is there that I born.

III

Iona: Mabouya Valley

St. Lucian conte, *or narrative Creole song, heard on the back
of an open truck travelling to Vieuxfort, some years ago*

Ma Kilman, Bon Dieu kai punir 'ous,
Pour qui raison parcequ' ous entrer trop religion.
Oui, l'autre coté, Bon Dieu kai benir 'ous,
Bon Dieu kai benir 'ous parcequi 'ous faire charité l'argent.

Corbeau aille Curaçao,
i' voyait l'argent ba 'ous,
ous prend l'argent cela,
ous mettait lui en cabaret.
Ous pas ka lire, ecrire, 'ous pas ka parler Anglais,

ous tait supposer ca; cabaret pas ni benefice.
L'heure Corbeau devirait,
l'tait ni, l' tait ni l'argent,
L'heure i' rivait ici.
Oui, maman! Corbeau kai fou!

Iona dit Corbeau, pendant 'ous tait Curaçao,
Moi fait deux 'tits mamaille, venir garder si c'est ca 'ous.
Corbeau criait: "Mama! Bonsoir, messieurs, mesdames,
lumer lampe-la ba moi,
pour moi garder ces mamailles-la!"
Corbeau virait dire: "Moi save toutes negres ka semble.
i peut si pas ca moin,
moi kai soigner ces mamailles-la!"

Oui, Corbeau partit, Corbeau descend Roseau,
allait chercher travail, pourqui 'peut soigner ces mamailles-la.
Iona dit Corbeau, pas tait descendre Roseau.

Mais i' descend Roseau, jamettes Roseau tomber derriere-i'.
Philippe Mago achetait un sax bai Corbeau,
i' pas ni temps jouer sax-la,
saxman comme lui prendre la vie-lui.

Samedi bon matin, Corbeau partit descendre en ville.
Samedi apres-midi, nous 'tendre la mort Corbeau.
Ca fait moi la peine, oui, ca brulait coeur-moin;
ca penetrait moin, l'heure moin 'tendre la mort Corbeau.

Iona dit comme-ca: ca qui fait lui la peine,
ca qui brulait coeur-lui, saxophone Corbeau pas jouer.

Moin 'tendre un corne cornait
a sur bord roseaux-a.
Moi dit: "Doux-doux, moin kai chercher
volants ba 'ous."
L'heure moin 'rivait la, moin fait raconte epi Corbeau.
I' dit: "Corne-la qui cornait-a,
c'est Iona ka cornait moin."

Guitar man la ka dire:
"Nous tous les deux c'est guitar man,
pas prendre ca pour un rien,
c'est meme beat-la nous ka chember."

Iona mariee, Dimanche a quatre heures.
Mardi, a huit heures, i' aille l'hopital.
I'fait un bombe, mari-lui cassait bras-lui.
L'heure moi joindre maman-ous,
Moin kai conter toute ca 'ous 'ja faire moin.
Iona!
(N'ai dit maman-ous!)
Iona!
(Ous pas ka 'couter moin!)

Trois jours, trois nuits
Iona bouillit, Iona pas chuitte.
(N'ai dit maman-i' ca.)
Toute moune ka dit Iona tourner,
C'est pas tourner Iona tourner,
mauvais i' mauvais,
Iona!

IV

Iona: Mabouya Valley

[*for Eric Branford*]

Ma Kilman, God will punish you,
for the reason that you've got too much religion.
On the other hand, God will bless you,
God will bless you because of your charity.

Corbeau went to Curaçao,
he sent you money back,
you took the same money
and put it in a rumshop.
You can't read, you can't write, you can't speak English,
you should know that rumshops make no profit.
When Corbeau came back,
he had, yes he had money,
when he arrived back here.
Yes, Mama, Corbeau'll go crazy!

Iona told Corbeau, while you were in Curaçao,
I made two little children, come and see if they're yours.
Corbeau cried out, "Mama! Good night, ladies and gentlemen,
light the lamp there for me,
for me to look at these kids!"
Corbeau came back and said, "I know niggers resemble,
they may or may not be mine,
I'll mind them all the same!"

Ah yes, Corbeau then left, he went down to Roseau,
he went to look for work, to mind the two little ones.
Iona told Corbeau, don't go down to Roseau.

But he went to Roseau, and Roseau's whores fell on him.
Philippe Mago brought Corbeau a saxophone,
he had no time to play the sax,
a saxman just like him took away his living.

Saturday morning early, Corbeau goes into town.
Saturday afternoon we hear Corbeau is dead.
That really made me sad, that really burnt my heart;
that really went through me when I heard Corbeau was dead.

Iona said like this: it made her sorry too,
it really burnt her heart, that the saxophone will never play.

I heard a horn blowing
by the river reeds down there.
"Sweetheart," I said, "I'll go looking
for flying fish for you."
When I got there, I came across Corbeau.
He said: "That horn you heard
was Iona horning me."

The guitar man's saying:
"We both are guitar men,
don't take it for anything,
we both holding the same beat."

Iona got married, Sunday at four o'clock.
Tuesday, by eight o'clock, she's in the hospital.

She made a fare, her husband broke her arm.
When I meet your mother,
I'll tell what you did me.
Iona!
(I'll tell your mama!)
Iona!
(You don't listen to me!)

Three days and three nights
Iona boiled, she's still not cooked.
(I'll tell her mother that.)
They say Iona's changed,
it isn't changed Iona's changed,
she's wicked, wicked, that's all,
Iona!

V

*For the Altarpiece of the Roseau Valley Church,
St. Lucia*

1

The chapel, as the pivot of this valley,
round which whatever is rooted loosely turns—
men, women, ditches, the revolving fields
of bananas, the secondary roads—
draws all to it, to the altar
and the massive altarpiece,
like a dull mirror, life
repeated there,
the common life outside
and the other life it holds;
a good man made it.

Two earth-brown labourers
dance the botay in it, the drum sounds under
the earth, the heavy foot.

                This is a rich valley,
It is fat with things.
Its roads radiate like aisles from the altar towards
those acres of bananas, towards
leaf-crowded mountains
rain-bellied clouds
in haze, in iron heat;

              This is a cursed valley,
ask the broken mules, the swollen children,
ask the dried women, their gap-toothed men,
ask the parish priest, who, in the altarpiece,
carries a replica of the church,
ask the two who could be Eve and Adam dancing.

  2
Five centuries ago
in the time of Giotto
this altar might have had
in one corner, when God was young,
St. Omer me fecit aetat, whatever his own age now,
Gloria Dei and to God's Mother also.

It is signed with music.
It turns the whole island.
You have to imagine it empty on a Sunday afternoon
between adorations

Nobody can see it and it is there,
nobody adores the two who could be Eve and Adam dancing.

A Sunday at three o'clock
when the real Adam and Eve have coupled
and lie in rechristening sweat

his sweat on her still breasts,
her sweat on his panelled torso
that hefts bananas
that has killed snakes
that has climbed out of rivers,

now, as on the furred tops of the hills,
a breeze moving the hairs on his chest
on a Sunday at three o'clock
when the snake pours itself
into a chalice of leaves.

The sugar factory is empty.

Nobody picks bananas,
no trucks raising dust on their way to Vieuxfort,
no helicopter spraying

the mosquito's banjo, yes,
and the gnat's violin, okay,

okay, not absolute Adamic silence,
the valley of Roseau is not the Garden of Eden,
and those who inhabit it are not in heaven,

so there are little wires of music
some marron up in the hills, by Aux Lyons,
some christening.

A boy banging a tin by the river,
with the river trying to sleep.
But nothing can break that silence,

which comes from the depth of the world,
from whatever one man believes he knows of God
and the suffering of his kind,

it comes from the wall of the altarpiece
ST. OMER AD GLORIAM DEI FECIT
in whatever year of his suffering.

    3

After so many bottles of white rum in a pile,
after the flight of so many little fishes
from the brush that is the finger of St. Francis,

after the deaths
of as many names as you want,
Iona, Julian, Ti-Nomme, Cacao,
like the death of the cane crop in Roseau Valley, St. Lucia.

After five thousand novenas
and the idea of the Virgin
coming and going like a little lamp,

after all that,
your faith like a canoe at evening coming in,

like a relative who is tired of America,
like a woman coming back to your house,

that sang in the ropes of your wrist
when you lifted this up;
so that, from time to time, on Sundays

between adorations, one might see,
if one were there, and not there,
looking in at the windows

the real faces of angels.

Volcano

Joyce was afraid of thunder,
but lions roared at his funeral
from the Zurich zoo.
Was it Zurich or Trieste?
No matter. These are legends, as much
as the death of Joyce is a legend,
or the strong rumour that Conrad
is dead, and that *Victory* is ironic.
On the edge of the night-horizon
from this beach house on the cliffs
there are now, till dawn,
two glares from the miles-out-
at-sea derricks; they are like
the glow of the cigar
and the glow of the volcano
at *Victory*'s end.
One could abandon writing
for the slow-burning signals
of the great, to be, instead,
their ideal reader, ruminative,
voracious, making the love of masterpieces
superior to attempting
to repeat or outdo them,
and be the greatest reader in the world.
At least it requires awe,

which has been lost to our time;
so many people have seen everything,
so many people can predict,
so many refuse to enter the silence
of victory, the indolence
that burns at the core,
so many are no more than
erect ash, like the cigar,
so many take thunder for granted.
How common is the lightning,
how lost the leviathans
we no longer look for!
There were giants in those days.
In those days they made good cigars.
I must read more carefully.

Endings

Things do not explode,
they fail, they fade,

as sunlight fades from the flesh,
as the foam drains quick in the sand,

even love's lightning flash
has no thunderous end,

it dies with the sound
of flowers fading like the flesh

from sweating pumice stone,
everything shapes this

till we are left
with the silence that surrounds Beethoven's head.

The Fist

The fist clenched round my heart
loosens a little, and I gasp
brightness; but it tightens
again. When have I ever not loved
the pain of love? But this has moved

past love to mania. This has the strong
clench of the madman, this is
gripping the ledge of unreason, before
plunging howling into the abyss.

Hold hard then, heart. This way at least you live.

Love after Love

The time will come
when, with elation,
you will greet yourself arriving
at your own door, in your own mirror,
and each will smile at the other's welcome,

and say, sit here. Eat.
You will love again the stranger who was your self.
Give wine. Give bread. Give back your heart
to itself, to the stranger who has loved you

all your life, whom you ignored
for another, who knows you by heart.
Take down the love letters from the bookshelf,

the photographs, the desperate notes,
peel your own image from the mirror.
Sit. Feast on your life.

Dark August

So much rain, so much life like the swollen sky
of this black August. My sister, the sun,
broods in her yellow room and won't come out.

Everything goes to hell; the mountains fume
like a kettle, rivers overrun; still,
she will not rise and turn off the rain.

She's in her room, fondling old things,
my poems, turning her album. Even if thunder falls
like a crash of plates from the sky,

she does not come out.
Don't you know I love you but am hopeless
at fixing the rain? But I am learning slowly

to love the dark days, the steaming hills,
the air with gossiping mosquitoes,
and to sip the medicine of bitterness,

so that when you emerge, my sister,
parting the beads of the rain,
with your forehead of flowers and eyes of forgiveness,

all will not be as it was, but it will be true
(you see they will not let me love
as I want), because, my sister, then

I would have learnt to love black days like bright ones,
the black rain, the white hills, when once
I loved only my happiness and you.

Sea Canes

Half my friends are dead.
I will make you new ones, said earth.
No, give me them back, as they were, instead,
with faults and all, I cried.

Tonight I can snatch their talk
from the faint surf's drone
through the canes, but I cannot walk

on the moonlit leaves of ocean
down that white road alone,
or float with the dreaming motion

of owls leaving earth's load.
O earth, the number of friends you keep
exceeds those left to be loved.

The sea canes by the cliff flash green and silver;
they were the seraph lances of my faith,
but out of what is lost grows something stronger

that has the rational radiance of stone,
enduring moonlight, further than despair,
strong as the wind, that through dividing canes

brings those we love before us, as they were,
with faults and all, not nobler, just there.

Midsummer, Tobago

Broad sun-stoned beaches.

White heat.
A green river.

A bridge,
scorched yellow palms

from the summer-sleeping house
drowsing through August.

Days I have held,
days I have lost,

days that outgrow, like daughters,
my harbouring arms.

Oddjob, a Bull Terrier

You prepare for one sorrow,
but another comes.
It is not like the weather,
you cannot brace yourself,
the unreadiness is all.
Your companion, the woman,
the friend next to you,
the child at your side,
and the dog,
we tremble for them,
we look seaward and muse
it will rain.
We shall get ready for rain;
you do not connect
the sunlight altering
the darkening oleanders
in the sea-garden,
the gold going out of the palms.
You do not connect this,
the fleck of the drizzle
on your flesh,
with the dog's whimper,
the thunder doesn't frighten,
the readiness is all;
what follows at your feet

is trying to tell you
the silence is all:
it is deeper than the readiness,
it is sea-deep,
earth-deep,
love-deep.

The silence
is stronger than thunder,
we are stricken dumb and deep
as the animals who never utter love
as we do, except
it becomes unutterable
and must be said,
in a whimper,
in tears,
in the drizzle that comes to our eyes
not uttering the loved thing's name,
the silence of the dead,
the silence of the deepest buried love is
the one silence,
and whether we bear it for beast,
for child, for woman, or friend,
it is the one love, it is the same,
and it is blest
deepest by loss
it is blest, it is blest.

Winding Up

I live on the water,
alone. Without wife and children.
I have circled every possibility
to come to this:

a low house by grey water,
with windows always open
to the stale sea. We do not choose such things,

but we are what we have made.
We suffer, the years pass,
we shed freight but not our need

for encumbrances. Love is a stone
that settled on the seabed
under grey water. Now, I require nothing

from poetry but true feeling,
no pity, no fame, no healing. Silent wife,
we can sit watching grey water,

and in a life awash
with mediocrity and trash
live rock-like.

I shall unlearn feeling,
unlearn my gift. That is greater
and harder than what passes there for life.

The Morning Moon

Still haunted by the cycle of the moon
racing full sail
past the crouched whale's back of Morne Coco Mountain,

I gasp at her sane brightness.

It's early December,
the breeze freshens the skin of this earth,
the goose-skin of water,

and I notice the blue plunge
of shadows down Morne Coco Mountain,
December's sundial,

happy that the earth is still changing,
that the full moon can blind me with her forehead
this bright foreday morning,

and that fine sprigs of white are springing from my beard.

To Return to the Trees

[for John Figueroa]

Senex, an oak.
Senex, this old sea-almond
unwincing in spray

in this geriatric grove
on the sea-road to Cumana.
To return to the trees,

to decline like this tree,
the burly oak
of Boanerges Ben Jonson!

Or am I lying
like this felled almond
when I write I look forward to age—

a gnarled poet
bearded with the whirlwind,
his metres like thunder?

It is not only the sea,
no, for on windy, green mornings
I read the changes on Morne Coco Mountain,

from flagrant sunrise
to its ashen end;
grey has grown strong to me,

it's no longer neutral,
no longer the dirty flag
of courage going under,

it is speckled with hues
like quartz, it's as
various as boredom,

grey now is a crystal
haze, a dull diamond,
stone-dusted and stoic,

grey is the heart at peace,
tougher than the warrior
as it bestrides factions,

it is the great pause
when the pillars of the temple
rest on Samson's palms

and are held, held,
that moment
when the heavy rock of the world

like a child sleeps
on the trembling shoulders of Atlas
and his own eyes close,

the toil that is balance.
Seneca, that fabled bore,
and his gnarled, laborious Latin

I can read only in fragments
of broken bark, his
heroes tempered by whirlwinds,

who see with the word
"senex," with its two eyes,
through the boles of this tree,

beyond joy,
beyond lyrical utterance,
this obdurate almond

going under the sand
with this language, slowly,
by sand grains, by centuries.

From

THE STAR-APPLE

KINGDOM

[1979]

The Schooner Flight

1 *Adios, Carenage*

In idle August, while the sea soft,
and leaves of brown islands stick to the rim
of this Caribbean, I blow out the light
by the dreamless face of Maria Concepcion
to ship as a seaman on the schooner *Flight*.
Out in the yard turning grey in the dawn,
I stood like a stone and nothing else move
but the cold sea rippling like galvanize
and the nail holes of stars in the sky roof,
till a wind start to interfere with the trees.
I pass me dry neighbour sweeping she yard
as I went downhill, and I nearly said:
"Sweep soft, you witch, 'cause she don't sleep hard,"
but the bitch look through me like I was dead.
A route taxi pull up, park-lights still on.
The driver size up my bags with a grin:
"This time, Shabine, like you really gone!"
I ain't answer the ass, I simply pile in
the back seat and watch the sky burn
above Laventille pink as the gown
in which the woman I left was sleeping,
and I look in the rearview and see a man
exactly like me, and the man was weeping
for the houses, the streets, that whole fucking island.

Christ have mercy on all sleeping things!
From that dog rotting down Wrightson Road
to when I was a dog on these streets;
if loving these islands must be my load,
out of corruption my soul takes wings,
But they had started to poison my soul
with their big house, big car, big-time bohbohl,
coolie, nigger, Syrian, and French Creole,
so I leave it for them and their carnival—
I taking a sea-bath, I gone down the road.
I know these islands from Monos to Nassau,
a rusty head sailor with sea-green eyes
that they nickname Shabine, the patois for
any red nigger, and I, Shabine, saw
when these slums of empire was paradise.
I'm just a red nigger who love the sea,
I had a sound colonial education,
I have Dutch, nigger, and English in me,
and either I'm nobody, or I'm a nation.

But Maria Concepcion was all my thought
watching the sea heaving up and down
as the port side of dories, schooners, and yachts
was painted afresh by the strokes of the sun
signing her name with every reflection;
I knew when dark-haired evening put on
her bright silk at sunset, and, folding the sea,
sidled under the sheet with her starry laugh,
that there'd be no rest, there'd be no forgetting.
Is like telling mourners round the graveside
about resurrection, they want the dead back,

so I smile to myself as the bow rope untied
and the *Flight* swing seaward: "Is no use repeating
that the sea have more fish. I ain't want her
dressed in the sexless light of a seraph,
I want those round brown eyes like a marmoset, and
till the day when I can lean back and laugh,
those claws that tickled my back on sweating
Sunday afternoons, like a crab on wet sand."
As I worked, watching the rotting waves come
past the bow that scissor the sea like silk,
I swear to you all, by my mother's milk,
by the stars that shall fly from tonight's furnace,
that I loved them, my children, my wife, my home;
I loved them as poets love the poetry
that kills them, as drowned sailors the sea.

You ever look up from some lonely beach
and see a far schooner? Well, when I write
this poem, each phrase go be soaked in salt;
I go draw and knot every line as tight
as ropes in this rigging; in simple speech
my common language go be the wind,
my pages the sails of the schooner *Flight*.
But let me tell you how this business begin.

2   *Raptures of the Deep*

Smuggled Scotch for O'Hara, big government man,
between Cedros and the Main, so the Coast Guard couldn't
    touch us,
and the Spanish pirogues always met us halfway,

but a voice kept saying: "Shabine, see this business
of playing pirate?" Well, so said, so done!
That whole racket crash. And I for a woman,
for her laces and silks, Maria Concepcion.
Ay, ay! Next thing I hear, some Commission of Enquiry
was being organized to conduct a big quiz,
with himself as chairman investigating himself.
Well, I knew damn well who the suckers would be,
not that shark in shark skin, but his pilot fish,
khaki-pants red niggers like you and me.
What worse, I fighting with Maria Concepcion,
plates flying and thing, so I swear: "Not again!"
It was mashing up my house and my family.
I was so broke all I needed was shades and a cup
or four shades and four cups in four-cup Port of Spain;
all the silver I had was the coins on the sea.

You saw them ministers in *The Express*,
guardians of the poor—one hand at their back,
and one set o' police only guarding their house,
and the Scotch pouring in through the back door.
As for that minister-monster who smuggled the booze,
that half-Syrian saurian, I got so vex to see
that face thick with powder, the warts, the stone lids
like a dinosaur caked with primordial ooze
by the lightning of flashbulbs sinking in wealth,
that I said: "Shabine, this is shit, understand!"
But he get somebody to kick my crutch out his office
like I was some artist! That bitch was so grand,
couldn't get off his high horse and kick me himself.
I have seen things that would make a slave sick
in this Trinidad, the Limers' Republic.

I couldn't shake the sea noise out of my head,
the shell of my ears sang Maria Concepcion,
so I start salvage diving with a crazy Mick,
name O'Shaughnessy, and a limey named Head;
but this Caribbean so choke with the dead
that when I would melt in emerald water,
whose ceiling rippled like a silk tent,
I saw them corals: brain, fire, sea-fans,
dead-men's-fingers, and then, the dead men.
I saw that the powdery sand was their bones
ground white from Senegal to San Salvador,
so, I panic third dive, and surface for a month
in the Seamen's Hostel. Fish broth and sermons.
When I thought of the woe I had brought my wife,
when I saw my worries with that other woman,
I wept under water, salt seeking salt,
for her beauty had fallen on me like a sword
cleaving me from my children, flesh of my flesh!

There was this barge from St. Vincent, but she was too deep
to float her again. When we drank, the limey
got tired of my sobbing for Maria Concepcion.
He said he was getting the bends. Good for him!
The pain in my heart for Maria Concepcion,
the hurt I had done to my wife and children,
was worse than the bends. In the rapturous deep
there was no cleft rock where my soul could hide
like the boobies each sunset, no sandbar of light
where I could rest, like the pelicans know,
so I got raptures once, and I saw God
like a harpooned grouper bleeding, and a far
voice was rumbling, "Shabine, if you leave her,

if you leave her, I shall give you the morning star."
When I left the madhouse I tried other women
but, once they stripped naked, their spiky cunts
bristled like sea-eggs and I couldn't dive.
The chaplain came round. I paid him no mind.
Where is my rest place, Jesus? Where is my harbour?
Where is the pillow I will not have to pay for,
and the window I can look from that frames my life?

3 Shabine Leaves the Republic

I had no nation now but the imagination.
After the white man, the niggers didn't want me
when the power swing to their side.
The first chain my hands and apologize, "History";
the next said I wasn't black enough for their pride.
Tell me, what power, on these unknown rocks—
a spray-plane Air Force, the Fire Brigade,
the Red Cross, the Regiment, two, three police dogs
that pass before you finish bawling "Parade!"?
I met History once, but he ain't recognize me,
a parchment Creole, with warts
like an old sea-bottle, crawling like a crab
through the holes of shadow cast by the net
of a grille balcony; cream linen, cream hat.
I confront him and shout, "Sir, is Shabine!
They say I'se your grandson. You remember Grandma,
your black cook, at all?" The bitch hawk and spat.
A spit like that worth any number of words.
But that's all them bastards have left us: words.

I no longer believed in the revolution.
I was losing faith in the love of my woman.
I had seen that moment Aleksandr Blok
crystallize in *The Twelve*. Was between
the Police Marine Branch and Hotel Venezuelana
one Sunday at noon. Young men without flags
using shirts, their chests waiting for holes.
They kept marching into the mountains, and
their noise ceased as foam sinks into sand.
They sank in the bright hills like rain, every one
with his own nimbus, leaving shirts in the street,
and the echo of power at the end of the street.
Propeller-blade fans turn over the Senate;
the judges, they say, still sweat in carmine,
on Frederick Street the idlers all marching
by standing still, the Budget turns a new leaf.
In the 12:30 movies the projectors best
not break down, or you go see revolution. Aleksandr Blok
enters and sits in the third row of pit eating choc-
olate cone, waiting for a spaghetti Western with Clint Eastwood and featuring Lee Van Cleef.

    4   *The* Flight, *Passing Blanchisseuse*

Dusk. The *Flight* passing Blanchisseuse.
Gulls wheel like from a gun again,
and foam gone amber that was white,
lighthouse and star start making friends,
down every beach the long day ends,
and there, on that last stretch of sand,
on a beach bare of all but light,

dark hands start pulling in the seine
of the dark sea, deep, deep inland.

    5 *Shabine Encounters the Middle Passage*

Man, I brisk in the galley first thing next dawn,
brewing li'l coffee; fog coil from the sea
like the kettle steaming when I put it down
slow, slow, 'cause I couldn't believe what I see:
where the horizon was one silver haze,
the fog swirl and swell into sails, so close
that I saw it was sails, my hair grip my skull,
it was horrors, but it was beautiful.
We float through a rustling forest of ships
with sails dry like paper, behind the glass
I saw men with rusty eyeholes like cannons,
and whenever their half-naked crews cross the sun,
right through their tissue, you traced their bones
like leaves against the sunlight; frigates, barkentines,
the backward-moving current swept them on,
and high on their decks I saw great admirals,
Rodney, Nelson, de Grasse, I heard the hoarse orders
they gave those Shabines, and the forest
of masts sail right through the *Flight*,
and all you could hear was the ghostly sound
of waves rustling like grass in a low wind
and the hissing weeds they trailed from the stern;
slowly they heaved past from east to west
like this round world was some cranked water wheel,
every ship pouring like a wooden bucket
dredged from the deep; my memory revolve

on all sailors before me, then the sun
heat the horizon's ring and they was mist.

Next we pass slave ships. Flags of all nations,
our fathers below deck too deep, I suppose,
to hear us shouting. So we stop shouting. Who knows
who his grandfather is, much less his name?
Tomorrow our landfall will be the Barbados.

6   The Sailor Sings Back to the Casuarinas

You see them on the low hills of Barbados
bracing like windbreaks, needles for hurricanes,
trailing, like masts, the cirrus of torn sails;
when I was green like them, I used to think
those cypresses, leaning against the sea,
that take the sea-noise up into their branches,
are not real cypresses but casuarinas.
Now captain just call them Canadian cedars.
But cedars, cypresses, or casuarinas,
whoever called them so had a good cause,
watching their bending bodies wail like women
after a storm, when some schooner came home
with news of one more sailor drowned again.
Once the sound "cypress" used to make more sense
than the green "casuarinas," though, to the wind
whatever grief bent them was all the same,
since they were trees with nothing else in mind
but heavenly leaping or to guard a grave;
but we live like our names and you would have
to be colonial to know the difference,

to know the pain of history words contain,
to love those trees with an inferior love,
and to believe: "Those casuarinas bend
like cypresses, their hair hangs down in rain
like sailors' wives. They're classic trees, and we,
if we live like the names our masters please,
by careful mimicry might become men."

7   The Flight *Anchors in Castries Harbor*

When the stars self were young over Castries,
I loved you alone and I loved the whole world.
What does it matter that our lives are different?
Burdened with the loves of our different children?
When I think of your young face washed by the wind
and your voice that chuckles in the slap of the sea?
The lights are out on La Toc promontory,
except for the hospital. Across at Vigie
the marina arcs keep vigil. I have kept my own
promise, to leave you the one thing I own,
you whom I loved first: my poetry.
We here for one night. Tomorrow, the *Flight* will be gone.

8   *Fight with the Crew*

It had one bitch on board, like he had me mark—
that was the cook, some Vincentian arse
with a skin like a gommier tree, red peeling bark,
and wash-out blue eyes; he wouldn't give me a ease,
like he feel he was white. Had an exercise book,
this same one here, that I was using to write
my poetry, so one day this man snatch it

from my hand, and start throwing it left and right
to the rest of the crew, bawling out, "Catch it,"
and start mincing me like I was some hen
because of the poems. Some case is for fist,
some case is for tholing pin, some is for knife—
this one was for knife. Well, I beg him first,
but he keep reading, "O my children, my wife,"
and playing he crying, to make the crew laugh;
it move like a flying fish, the silver knife
that catch him right in the plump of his calf,
and he faint so slowly, and he turn more white
than he thought he was. I suppose among men
you need that sort of thing. It ain't right
but that's how it is. There wasn't much pain,
just plenty blood, and Vincie and me best friend,
but none of them go fuck with my poetry again.

9 Maria Concepcion & the Book of Dreams

The jet that was screeching over the *Flight*
was opening a curtain into the past.
"Dominica ahead!"
                          "It still have Caribs there."
"One day go be planes only, no more boat."
"Vince, God ain't make nigger to fly through the air."
"Progress, Shabine, that's what it's all about.
Progress leaving all we small islands behind."
I was at the wheel, Vince sitting next to me
gaffing. Crisp, bracing day. A high-running sea.
"Progress is something to ask Caribs about.
They kill them by millions, some in war,
some by forced labour dying in the mines

looking for silver, after that niggers; more
progress. Until I see definite signs
that mankind change, Vince, I ain't want to hear.
Progress is history's dirty joke.
Ask that sad green island getting nearer."
Green islands, like mangoes pickled in brine.
In such fierce salt let my wound be healed,
me, in my freshness as a seafarer.

That night, with the sky sparks frosty with fire,
I ran like a Carib through Dominica,
my nose holes choked with memory of smoke;
I heard the screams of my burning children,
I ate the brains of mushrooms, the fungi
of devil's parasols under white, leprous rocks;
my breakfast was leaf mould in leaking forests,
with leaves big as maps, and when I heard noise
of the soldiers' progress through the thick leaves,
though my heart was bursting, I get up and ran
through the blades of balisier sharper than spears;
with the blood of my race, I ran, boy, I ran
with moss-footed speed like a painted bird;
then I fall, but I fall by an icy stream under
cool fountains of fern, and a screaming parrot
catch the dry branches and I drowned at last
in big breakers of smoke; then when that ocean
of black smoke pass, and the sky turn white,
there was nothing but Progress, if Progress is
an iguana as still as a young leaf in sunlight.
I bawl for Maria, and her *Book of Dreams*.

It anchored her sleep, that insomniac's Bible,
a soiled orange booklet with a cyclops' eye
center, from the Dominican Republic.
Its coarse pages were black with the usual
symbols of prophecy, in excited Spanish;
an open palm upright, sectioned and numbered
like a butcher chart, delivered the future.
One night, in a fever, radiantly ill,
she say, "Bring me the book, the end has come."
She said: "I dreamt of whales and a storm,"
but for that dream, the book had no answer.
A next night I dreamed of three old women
featureless as silkworms, stitching my fate,
and I scream at them to come out my house,
and I try beating them away with a broom,
but as they go out, so they crawl back again,
until I start screaming and crying, my flesh
raining with sweat, and she ravage the book
for the dream meaning, and there was nothing;
my nerves melt like a jellyfish—that was when I broke—
they found me round the Savannah, screaming:

All you see me talking to the wind, so you think I mad.
Well, Shabine has bridled the horses of the sea;
you see me watching the sun till my eyeballs seared,
so all you mad people feel Shabine crazy,
but all you ain't know my strength, hear? The coconuts
standing by in their regiments in yellow khaki,
they waiting for Shabine to take over these islands,
and all you best dread the day I am healed
of being a human. All you fate in my hand,
ministers, businessmen, Shabine have you, friend,

I shall scatter your lives like a handful of sand,
I who have no weapon but poetry and
the lances of palms and the sea's shining shield!

    10   *Out of the Depths*

Next day, dark sea. A arse-aching dawn.
"Damn wind shift sudden as a woman mind."
The slow swell start cresting like some mountain range
with snow on the top.
                    "Ay, Skipper, sky dark!"
"This ain't right for August."
                    "This light damn strange,
this season, sky should be clear as a field."

A stingray steeplechase across the sea,
tail whipping water, the high man-o'-wars
start reeling inland, quick, quick an archery
of flying fish miss us! Vince say: "You notice?"
and a black-mane squall pounce on the sail
like a dog on a pigeon, and it snap the neck
of the *Flight* and shake it from head to tail.
"Be Jesus, I never see sea get so rough
so fast! That wind come from God back pocket!"
"Where Cap'n headin? Like the man gone blind!"
"If we's to drong, we go drong, Vince, fock-it!"
"Shabine, say your prayers, if life leave you any!"

I have not loved those that I loved enough.
Worse than the mule kick of Kick-'Em-Jenny
Channel, rain start to pelt the *Flight* between
mountains of water. If I was frighten?

The tent poles of water spouts bracing the sky
start wobbling, clouds unstitch at the seams
and sky water drench us, and I hear myself cry,
"I'm the drowned sailor in her *Book of Dreams*."
I remembered them ghost ships, I saw me corkscrewing
to the sea-bed of sea-worms, fathom pass fathom,
my jaw clench like a fist, and only one thing
hold me, trembling, how my family safe home.
Then a strength like it seize me and the strength said:
"I from backward people who still fear God."
Let Him, in His might, heave Leviathan upward
by the winch of His will, the beast pouring lace
from his sea-bottom bed; and that was the faith
that had fade from a child in the Methodist chapel
in Chisel Street, Castries, when the whale-bell
sang service and, in hard pews ribbed like the whale,
proud with despair, we sang how our race
survive the sea's maw, our history, our peril,
and now I was ready for whatever death will.
But if that storm had strength, was in Cap'n face,
beard beading with spray, tears salting the eyes,
crucify to his post, that nigger hold fast
to that wheel, man, like the cross held Jesus,
and the wounds of his eyes like they crying for us,
and I feeding him white rum, while every crest
with Leviathan-lash make the *Flight* quail
like two criminal. Whole night, with no rest,
till red-eyed like dawn, we watch our travail
subsiding, subside, and there was no more storm.
And the noon sea get calm as Thy Kingdom come.

11 *After the Storm*

There's a fresh light that follows a storm
while the whole sea still havoc; in its bright wake
I saw the veiled face of Maria Concepcion
marrying the ocean, then drifting away
in the widening lace of her bridal train
with white gulls her bridesmaids, till she was gone.
I wanted nothing after that day.
Across my own face, like the face of the sun,
a light rain was falling, with the sea calm.

Fall gently, rain, on the sea's upturned face
like a girl showering; make these islands fresh
as Shabine once knew them! Let every trace,
every hot road, smell like clothes she just press
and sprinkle with drizzle. I finish dream;
whatever the rain wash and the sun iron:
the white clouds, the sea and sky with one seam,
is clothes enough for my nakedness.
Though my *Flight* never pass the incoming tide
of this inland sea beyond the loud reefs
of the final Bahamas, I am satisfied
if my hand gave voice to one people's grief.
Open the map. More islands there, man,
than peas on a tin plate, all different size,
one thousand in the Bahamas alone,
from mountains to low scrub with coral keys,
and from this bowsprit, I bless every town,
the blue smell of smoke in hills behind them,
and the one small road winding down them like twine
to the roofs below; I have only one theme:

The bowsprit, the arrow, the longing, the lunging heart—
the flight to a target whose aim we'll never know,
vain search for one island that heals with its harbour
and a guiltless horizon, where the almond's shadow
doesn't injure the sand. There are so many islands!
As many islands as the stars at night
on that branched tree from which meteors are shaken
like falling fruit around the schooner *Flight*.
But things must fall, and so it always was,
on one hand Venus, on the other Mars;
fall, and are one, just as this earth is one
island in archipelagoes of stars.
My first friend was the sea. Now, is my last.
I stop talking now. I work, then I read,
cotching under a lantern hooked to the mast.
I try to forget what happiness was,
and when that don't work, I study the stars.
Sometimes is just me, and the soft-scissored foam
as the deck turn white and the moon open
a cloud like a door, and the light over me
is a road in white moonlight taking me home.
Shabine sang to you from the depths of the sea.

Sabbaths, W.I.

Those villages stricken with the melancholia of Sunday,
in all of whose ochre streets one dog is sleeping

those volcanoes like ashen roses, or the incurable sore
of poverty, around whose puckered mouth thin boys are
selling yellow sulphur stone

the burnt banana leaves that used to dance
the river whose bed is made of broken bottles
the cocoa grove where a bird whose cry sounds green and
yellow and in the lights under the leaves crested with
orange flame has forgotten its flute

gommiers peeling from sunburn still wrestling to escape the sea

the dead lizard turning blue as stone

those rivers, threads of spittle, that forgot the old music

that dry, brief esplanade under the drier sea almonds
where the dry old men sat

watching a white schooner stuck in the branches

and playing draughts with the moving frigate birds

those hillsides like broken pots
those ferns that stamped their skeletons on the skin

and those roads that begin reciting their names at vespers

mention them and they will stop
those crabs that were willing to let an epoch pass
those herons like spinsters that doubted their reflections
enquiring, enquiring

those nettles that waited
those Sundays, those Sundays

those Sundays when the lights at the road's end were an occasion

those Sundays when my mother lay on her back
those Sundays when the sisters gathered like white moths
round their street lantern

and cities passed us by on the horizon

The Sea Is History

Where are your monuments, your battles, martyrs?
Where is your tribal memory? Sirs,
in that grey vault. The sea. The sea
has locked them up. The sea is History.

First, there was the heaving oil,
heavy as chaos;
then, like a light at the end of a tunnel,

the lantern of a caravel,
and that was Genesis.
Then there were the packed cries,
the shit, the moaning:

Exodus.
Bone soldered by coral to bone,
mosaics
mantled by the benediction of the shark's shadow,

that was the Ark of the Covenant.
Then came from the plucked wires
of sunlight on the sea floor

the plangent harps of the Babylonian bondage,
as the white cowries clustered like manacles
on the drowned women,

and those were the ivory bracelets
of the Song of Solomon,
but the ocean kept turning blank pages

looking for History.
Then came the men with eyes heavy as anchors
who sank without tombs,

brigands who barbecued cattle,
leaving their charred ribs like palm leaves on the shore,
then the foaming, rabid maw

of the tidal wave swallowing Port Royal,
and that was Jonah,
but where is your Renaissance?

Sir, it is locked in them sea-sands
out there past the reef's moiling shelf,
where the men-o'-war floated down;

strop on these goggles, I'll guide you there myself.
It's all subtle and submarine,
through colonnades of coral,

past the gothic windows of sea-fans
to where the crusty grouper, onyx-eyed,
blinks, weighted by its jewels, like a bald queen;

and these groined caves with barnacles
pitted like stone
are our cathedrals,

and the furnace before the hurricanes:
Gomorrah. Bones ground by windmills
into marl and cornmeal,

and that was Lamentations—
that was just Lamentations,
it was not History;

then came, like scum on the river's drying lip,
the brown reeds of villages
mantling and congealing into towns,

and at evening, the midges' choirs,
and above them, the spires
lancing the side of God

as His son set, and that was the New Testament.

Then came the white sisters clapping
to the waves' progress,
and that was Emancipation—

jubilation, O jubilation—
vanishing swiftly
as the sea's lace dries in the sun,

but that was not History,
that was only faith,
and then each rock broke into its own nation;

then came the synod of flies,
then came the secretarial heron,
then came the bullfrog bellowing for a vote,

fireflies with bright ideas
and bats like jetting ambassadors
and the mantis, like khaki police,

and the furred caterpillars of judges
examining each case closely,
and then in the dark ears of ferns

and in the salt chuckle of rocks
with their sea pools, there was the sound
like a rumour without any echo

of History, really beginning.

Egypt, Tobago

[*for N.M.*]

There is a shattered palm
on this fierce shore,
its plumes the rusting helmet of a dead warrior.

Numb Antony, in the torpor
stretching her inert
sex near him like a sleeping cat,
knows his heart is the real desert.

Over the dunes
of her heaving,
to his heart's drumming
fades the mirage of the legions,

across love-tousled sheets,
the triremes fading.
At the carved door of her temple
a fly wrings its message.

He brushes a damp hair
away from an ear

as perfect as a sleeping child's.
He stares, inert, the fallen column.

He lies like a copper palm
tree at three in the afternoon
by a hot sea
and a river, in Egypt, Tobago.

Her salt marsh dries in the heat
where he foundered
without armour.
He exchanged an empire for her beads of sweat,

the uproar of arenas,
the changing surf
of senators, for
this silent ceiling over silent sand—

this grizzled bear, whose fur,
moulting, is silvered—
for this quick fox with her
sweet stench. By sleep dismembered,

his head
is in Egypt, his feet
in Rome, his groin a desert
trench with its dead soldier.

He drifts a finger
through her stiff hair
crisp as a mare's fountaining tail.
Shadows creep up the palace tile.

He is too tired to move;
a groan would waken
trumpets, one more gesture,
war. His glare,

a shield
reflecting fires,
a brass brow that cannot frown
at carnage, sweats the sun's force.

It is not the turmoil
of autumnal lust,
its treacheries, that drove
him, fired and grimed with dust,

this far, not even love,
but a great rage without
clamour, that grew great
because its depth is quiet;

it hears the river
of her young brown blood,
it feels the whole sky quiver
with her blue eyelid.

She sleeps with the soft engine of a child,

that sleep which scythes
the stalks of lances, fells the
harvest of legions
with nothing for its knives,
that makes Caesars,

sputtering at flies,
slapping their foreheads
with the laurel's imprint,
drunkards, comedians.

All-humbling sleep, whose peace
is sweet as death,
whose silence has
all the sea's weight and volubility,

who swings this globe by a hair's trembling breath.

Shattered and wild and
palm-crowned Antony,
rusting in Egypt,
ready to lose the world,
to Actium and sand,

everything else
is vanity, but this tenderness
for a woman not his mistress
but his sleeping child.

The sky is cloudless. The afternoon is mild.

The Saddhu of Couva

[*for Kenneth Ramchand*]

When sunset, a brass gong,
vibrate through Couva,
is then I see my soul, swiftly unsheathed,
like a white cattle bird growing more small
over the ocean of the evening canes,
and I sit quiet, waiting for it to return
like a hog-cattle blistered with mud,
because, for my spirit, India is too far.
And to that gong
sometimes bald clouds in saffron robes assemble
sacred to the evening,
sacred even to Ramlochan,
singing Indian hits from his jute hammock
while evening strokes the flanks
and silver horns of his maroon taxi,
as the mosquitoes whine their evening mantras,
my friend Anopheles, on the sitar,
and the fireflies making every dusk Divali.

I knot my head with a cloud,
my white moustache bristle like horns,
my hands are brittle as the pages of Ramayana.
Once the sacred monkeys multiplied like branches

in the ancient temples; I did not miss them,
because these fields sang of Bengal,
behind Ramlochan Repairs there was Uttar Pradesh;
but time roars in my ears like a river,
old age is a conflagration
as fierce as the cane fires of crop time.
I will pass through these people like a cloud,
they will see a white bird beating the evening sea
of the canes behind Couva,
and who will point it as my soul unsheathed?
Neither the bridegroom in beads,
nor the bride in her veils,
their sacred language on the cinema hoardings.

I talked too damn much on the Couva Village Council.
I talked too softly, I was always drowned
by the loudspeakers in front of the stores
or the loudspeakers with the greatest pictures.
I am best suited to stalk like a white cattle bird
on legs like sticks, with sticking to the Path
between the canes on a district road at dusk.
Playing the Elder. There are no more elders.
Is only old people.

My friends spit on the government.
I do not think is just the government.
Suppose all the gods too old,
Suppose they dead and they burning them,
supposing when some cane cutter
start chopping up snakes with a cutlass
he is severing the snake-armed god,
and suppose some hunter has caught

Hanuman in his mischief in a monkey cage.
Suppose all the gods were killed by electric light?

Sunset, a bonfire, roars in my ears;
embers of blown swallows dart and cry,
like women distracted,
around its cremation.
I ascend to my bed of sweet sandalwood.

Forest of Europe

[*for Joseph Brodsky*]

The last leaves fell like notes from a piano
and left their ovals echoing in the ear;
with gawky music stands, the winter forest
looks like an empty orchestra, its lines
ruled on these scattered manuscripts of snow.

The inlaid copper laurel of an oak
shines through the brown-bricked glass above your head
as bright as whisky, while the wintry breath
of lines from Mandelstam, which you recite,
uncoils as visibly as cigarette smoke.

"The rustling of ruble notes by the lemon Neva."
Under your exile's tongue, crisp under heel,
the gutturals crackle like decaying leaves,
the phrase from Mandelstam circles with light
in a brown room, in barren Oklahoma.

There is a Gulag Archipelago
under this ice, where the salt, mineral spring
of the long Trail of Tears runnels these plains
as hard and open as a herdsman's face
sun-cracked and stubbled with unshaven snow.

Growing in whispers from the Writers' Congress,
the snow circles like Cossacks round the corpse
of a tired Choctaw till it is a blizzard
of treaties and white papers as we lose
sight of the single human through the cause.

So every spring these branches load their shelves,
like libraries with newly published leaves,
till waste recycles them—paper to snow—
but, at zero of suffering, one mind
lasts like this oak with a few brazen leaves.

As the train passed the forest's tortured icons,
the floes clanging like freight yards, then the spires
of frozen tears, the station's screeching steam,
he drew them in a single winter's breath
whose freezing consonants turned into stones.

He saw the poetry in forlorn stations
under clouds vast as Asia, through districts
that could gulp Oklahoma like a grape,
not these tree-shaded prairie halts but space
so desolate it mocked destinations.

Who is that dark child on the parapets
of Europe, watching the evening river mint
its sovereigns stamped with power, not with poets,
the Thames and the Neva rustling like banknotes,
then, black on gold, the Hudson's silhouettes?

From frozen Neva to the Hudson pours,
under the airport domes, the echoing stations,

the tributary of emigrants whom exile
has made as classless as the common cold,
citizens of a language that is now yours,

and every February, every "last autumn,"
you write far from the threshing harvesters
folding wheat like a girl plaiting her hair,
far from Russia's canals quivering with sunstroke,
a man living with English in one room.

The tourist archipelagoes of my South
are prisons too, corruptible, and though
there is no harder prison than writing verse,
what's poetry, if it is worth its salt,
but a phrase men can pass from hand to mouth?

From hand to mouth, across the centuries,
the bread that lasts when systems have decayed,
when, in his forest of barbed-wire branches,
a prisoner circles, chewing the one phrase
whose music will last longer than the leaves,

whose condensation is the marble sweat
of angels' foreheads, which will never dry
till Borealis shuts the peacock lights
of its slow fan from L.A. to Archangel,
and memory needs nothing to repeat.

Frightened and starved, with divine fever
Osip Mandelstam shook, and every
metaphor shuddered him with ague,

each vowel heavier than a boundary stone,
"to the rustling of ruble notes by the lemon Neva,"

but now that fever is a fire whose glow
warms our hands, Joseph, as we grunt like primates
exchanging gutturals in this winter cave
of a brown cottage, while in drifts outside
mastodons force their systems through the snow.

Koenig of the River

Koenig knew now there was no one on the river.
Entering its brown mouth choking with lilies
and curtained with midges, Koenig poled the shallop
past the abandoned ferry and the ferry piles
coated with coal dust. Staying aboard, he saw, up
in a thick meadow, a sand-coloured mule,
untethered, with no harness, and no signs
of habitation round the ruined factory wheel
locked hard in rust, and through whose spokes the vines
of wild yam leaves leant from overweight;
the wild bananas in the yellowish sunlight
were dugged like aching cows with unmilked fruit.
This was the last of the productive mines.
Only the vegetation here looked right.
A crab of pain scuttled shooting up his foot
and fastened on his neck, at the brain's root.
He felt his reason curling back like parchment
in this fierce torpor. Well, he no longer taxed
and tired what was left of his memory;
he should thank heaven he had escaped the sea,
and anyway, he had demanded to be sent
here with the others—why get this river vexed
with his complaints? Koenig wanted to sing,
suddenly, if only to keep the river company—
this was a river, and Koenig, his name meant King.

They had all caught the missionary fever:
they were prepared to expiate the sins
of savages, to tame them as he would tame this river
subtly, as it flowed, accepting its bends;
he had seen how other missionaries met their ends—
swinging in the wind, like a dead clapper when
a bell is broken, if that sky was a bell—
for treating savages as if they were men,
and frightening them with talk of heaven and hell.
But I have forgotten our journey's origins,
mused Koenig, and our purpose. He knew it was noble,
based on some phrase, forgotten, from the Bible,
but he felt bodiless, like a man stumbling from
the pages of a novel, not a forest,
written a hundred years ago. He stroked his uniform,
clogged with the hooked burrs that had tried
to pull him, like the other drowning hands whom
his panic abandoned. The others had died,
like real men, by death. I, Koenig, am a ghost,
ghost-king of rivers. Well, even ghosts must rest.
If he knew he was lost he was not lost.
It was when you pretended that you were a fool.
He banked and leaned tiredly on the pole.
If I'm a character called Koenig, then I
shall dominate my future like a fiction
in which there is a real river and real sky,
so I'm not really tired, and should push on.

The lights between the leaves were beautiful,
and, as in that far life, now he was grateful
for any pool of light between the dull, usual
clouds of life: a sunspot haloed his tonsure;

silver and copper coins danced on the river;
his head felt warm—the light danced on his skull
like a benediction. Koenig closed his eyes,
and he felt blessed. It made direction sure.
He leant on the pole. He must push on some more.
He said his name. His voice sounded German,
then he said "river," but what was German
if he alone could hear it? *Ich spreche Deutsch*
sounded as genuine as his name in English,
Koenig in Deutsch, and, in English, King.
Did the river want to be called anything?
He asked the river. The river said nothing.

Around the bend the river poured its silver
like some remorseful mine, giving and giving
everything green and white: white sky, white
water, and the dull green like a drumbeat
of the slow-sliding forest, the green heat;
then, on some sandbar, a mirage ahead:
fabric of muslin sails, spiderweb rigging,
a schooner, foundered on black river mud,
was rising slowly up from the riverbed,
and a top-hatted native reading an inverted
newspaper.
          "Where's our Queen?" Koenig shouted.
"Where's our Kaiser?"
                    The nigger disappeared.
Koenig felt that he himself was being read
like the newspaper or a hundred-year-old novel.
"The Queen dead! Kaiser dead!" the voices shouted.
And it flashed through him those trunks were not wood
but that the ghosts of slaughtered Indians stood

there in the mangroves, their eyes like fireflies
in the green dark, and that like hummingbirds
they sailed rather than ran between the trees.
The river carried him past his shouted words.
The schooner had gone down without a trace.
"There was a time when we ruled everything,"
Koenig sang to his corrugated white reflection.
"The German Eagle and the British Lion,
we ruled worlds wider than this river flows,
worlds with dyed elephants, with tasselled howdahs,
tigers that carried the striped shade when they rose
from their palm coverts; men shall not see these days
again; our flags sank with the sunset on the dhows
of Egypt; we ruled rivers as huge as the Nile,
the Ganges, and the Congo, we tamed, we ruled
you when our empires reached their blazing peak."
This was a small creek somewhere in the world,
never mind where—victory was in sight.
Koenig laughed and spat in the brown creek.
The mosquitoes now were singing to the night
that rose up from the river, the fog uncurled
under the mangroves. Koenig clenched each fist
around his barge-pole sceptre, as a mist
rises from the river and the page goes white.

The Star-Apple Kingdom

There were still shards of an ancient pastoral
in those shires of the island where the cattle drank
their pools of shadow from an older sky,
surviving from when the landscape copied such subjects as
"Herefords at Sunset in the Valley of the Wye."
The mountain water that fell white from the mill wheel
sprinkling like petals from the star-apple trees,
and all of the windmills and sugar mills moved by mules
on the treadmill of Monday to Monday, would repeat
in tongues of water and wind and fire, in tongues
of Mission School pickaninnies, like rivers remembering
their source, Parish Trelawny, Parish St. David, Parish
St. Andrew, the names afflicting the pastures,
the lime groves and fences of marl stone and the cattle
with a docile longing, an epochal content.
And there were, like old wedding lace in an attic,
among the boas and parasols and the tea-coloured
daguerreotypes, hints of an epochal happiness
as ordered and infinite to the child
as the great house road to the Great House
down a perspective of casuarinas plunging green manes
in time to the horses, an orderly life
reduced by lorgnettes day and night, one disc the sun,
the other the moon, reduced into a pier glass:
nannies diminished to dolls, mahogany stairways

no larger than those of an album in which
the flash of cutlery yellows, as gamboge as
the piled cakes of teatime on that latticed
bougainvillea verandah that looked down toward
a prospect of Cuyp-like Herefords under a sky
lurid as a porcelain souvenir with these words:
"Herefords at Sunset in the Valley of the Wye."

Strange, that the rancour of hatred hid in that dream
of slow rivers and lily-like parasols, in snaps
of fine old colonial families, curled at the edge
not from age or from fire or the chemicals, no, not at all,
but because, off at its edges, innocently excluded
stood the groom, the cattle boy, the housemaid, the gardeners,
the tenants, the good Negroes down in the village,
their mouths in the locked jaw of a silent scream.
A scream which would open the doors to swing wildly
all night, that was bringing in heavier clouds,
more black smoke than cloud, frightening the cattle
in whose bulging eyes the Great House diminished;
a scorching wind of a scream
that began to extinguish the fireflies,
that dried the water mill creaking to a stop
as it was about to pronounce Parish Trelawny
all over, in the ancient pastoral voice,
a wind that blew all without bending anything,
neither the leaves of the album nor the lime groves;
blew Nanny floating back in white from a feather
to a chimerical, chemical pin speck that shrank
the drinking Herefords to brown porcelain cows
on a mantelpiece, Trelawny trembling with dusk,
the scorched pastures of the old benign Custos; blew

far the decent servants and the lifelong cook,
and shriveled to a shard that ancient pastoral
of dusk in a gilt-edged frame now catching the evening sun
in Jamaica, making both epochs one.

He looked out from the Great House windows on
clouds that still held the fragrance of fire,
he saw the Botanical Gardens officially drown
in a formal dusk, where governors had strolled
and black gardeners had smiled over glinting shears
at the lilies of parasols on the floating lawns,
the flame trees obeyed his will and lowered their wicks,
the flowers tightened their fists in the name of thrift,
the porcelain lamps of ripe cocoa, the magnolia's jet
dimmed on the one circuit with the ginger lilies
and left a lonely bulb on the verandah,
and, had his mandate extended to that ceiling
of star-apple candelabra, he would have ordered
the sky to sleep, saying, I'm tired,
save the starlight for victories, we can't afford it,
leave the moon on for one more hour, and that's it.
But though his power, the given mandate, extended
from tangerine daybreaks to star-apple dusks,
his hand could not dam that ceaseless torrent of dust
that carried the shacks of the poor, to their root-rock music,
down the gullies of Yallahs and August Town,
to lodge them on thorns of maca, with their rags
crucified by cactus, tins, old tires, cartons;
from the black Warieka Hills the sky glowed fierce as
the dials of a million radios,
a throbbing sunset that glowed like a grid
where the dread beat rose from the jukebox of Kingston.

He saw the fountains dried of quadrilles, the water-music
of the country dancers, the fiddlers like fifes
put aside. He had to heal
this malarial island in its bath of bay leaves,
its forests tossing with fever, the dry cattle
groaning like winches, the grass that kept shaking
its head to remember its name. No vowels left
in the mill wheel, the river. Rock stone. Rock stone.

The mountains rolled like whales through phosphorous stars,
as he swayed like a stone down fathoms into sleep,
drawn by that magnet which pulls down half the world
between a star and a star, by that black power
that has the assassin dreaming of snow,
that poleaxes the tyrant to a sleeping child.
The house is rocking at anchor, but as he falls
his mind is a mill wheel in moonlight,
and he hears, in the sleep of his moonlight, the drowned
bell of Port Royal's cathedral, sees the copper pennies
of bubbles rising from the empty eye-pockets
of green buccaneers, the parrot fish floating
from the frayed shoulders of pirates, sea-horses
drawing gowned ladies in their liquid promenade
across the moss-green meadows of the sea;
he heard the drowned choirs under Palisadoes,
a hymn ascending to earth from a heaven inverted
by water, a crab climbing the steeple,
and he climbed from that submarine kingdom
as the evening lights came on in the institute,
the scholars lamplit in their own aquarium,
he saw them mouthing like parrot fish, as he passed
upward from that baptism, their history lessons,

the bubbles like ideas which he could not break:
Jamaica was captured by Penn and Venables,
Port Royal perished in a cataclysmic earthquake.

Before the coruscating façades of cathedrals
from Santiago to Caracas, where the penitential archbishops
washed the feet of paupers (a parenthetical moment
that made the Caribbean a baptismal font,
turned butterflies to stone, and whitened like doves
the buzzards circling municipal garbage),
the Caribbean was borne like an elliptical basin
in the hands of acolytes, and a people were absolved
of a history which they did not commit;
the slave pardoned his whip, and the dispossessed
said the rosary of islands for three hundred years,
a hymn that resounded like the hum of the sea
inside a sea-cave, as their knees turned to stone,
while the bodies of patriots were melting down walls
still crusted with mute outcries of *La Revolución!*
"San Salvador, pray for us, St. Thomas, San Domingo,
ora pro nobis, intercede for us, Sancta Lucia
of no eyes," and when the circular chaplet
reached the last black bead of Sancta Trinidad
they began again, their knees drilled into stone,
where Colón had begun, with San Salvador's bead,
beads of black colonies round the necks of Indians.
And while they prayed for an economic miracle,
ulcers formed on the municipal portraits,
the hotels went up, and the casinos and brothels,
and the empires of tobacco, sugar, and bananas,
until a black woman, shawled like a buzzard,
climbed up the stairs and knocked at the door

of his dream, whispering in the ear of the keyhole:
"Let me in, I'm finished with praying, I'm the Revolution.
I am the darker, the older America."

She was as beautiful as a stone in the sunrise,
her voice had the gutturals of machine guns
across khaki deserts where the cactus flower
detonates like grenades, her sex was the slit throat
of an Indian, her hair had the blue-black sheen of the crow.
She was a black umbrella blown inside out
by the wind of revolution, La Madre Dolorosa,
a black rose of sorrow, a black mine of silence,
raped wife, empty mother, Aztec virgin
transfixed by arrows from a thousand guitars,
a stone full of silence, which, if it gave tongue
to the tortures done in the name of the Father,
would curdle the blood of the marauding wolf,
the fountain of generals, poets, and cripples
who danced without moving over their graves
with each revolution; her Caesarean was stitched
by the teeth of machine guns, and every sunset
she carried the Caribbean's elliptical basin
as she had once carried the penitential napkins
to be the footbath of dictators, Trujillo, Machado,
and those whose faces had yellowed like posters
on municipal walls. Now she stroked his hair
until it turned white, but she would not understand
that he wanted no other power but peace,
that he wanted a revolution without any bloodshed,
he wanted a history without any memory,
streets without statues,
and a geography without myth. He wanted no armies

but those regiments of bananas, thick lances of cane,
and he sobbed, "I am powerless, except for love."
She faded from him, because he could not kill;
she shrank to a bat that hung day and night
in the back of his brain. He rose in his dream.

The soul, which was his body made as thin
as its reflection and invulnerable
without its clock, was losing track of time;
it walked the mountain tracks of the Maroons,
it swung with Gordon from the creaking gibbet,
it bought a pack of peppermints and cashews
from one of the bandanna'd mammies outside the ward,
it heard his breath pitched to the decibels
of the peanut vendors' carts, it entered a municipal wall
stirring the slogans that shrieked his name: SAVIOUR!
and others: LACKEY! he melted like a spoon
through the alphabet soup of CIA, PNP, OPEC,
that resettled once he passed through with this thought:
I should have foreseen those seraphs with barbed-wire hair,
beards like burst mattresses, and wild eyes of garnet,
who nestled the Coptic Bible to their ribs, would
call me Joshua, expecting him to bring down Babylon
by Wednesday, after the fall of Jericho; yes, yes,
I should have seen the cunning bitterness of the rich
who left me no money but these mandates:

His aerial mandate, which
contained the crows whose circuit
was this wedding band that married him to his island.
His marine mandate, which
was the fishing limits

which the shark scissored like silk with its teeth
between Key West and Havana;
his terrestrial:
the bled hills rusted with bauxite;
paradisal:
the chimneys like angels sheathed in aluminium.

In shape like a cloud
he saw the face of his father,
the hair like white cirrus blown back
in a photographic wind,
the mouth of mahogany winced shut,
the eyes lidded, resigned
to the first compromise,
the last ultimatum,
the first and last referendum.

One morning the Caribbean was cut up
by seven prime ministers who bought the sea in bolts—
one thousand miles of aquamarine with lace trimmings,
one million yards of lime-coloured silk,
one mile of violet, leagues of cerulean satin—
who sold it at a markup to the conglomerates,
the same conglomerates who had rented the water spouts
for ninety-nine years in exchange for fifty ships,
who retailed it in turn to the ministers
with only one bank account, who then resold it
in ads for the Caribbean Economic Community,
till everyone owned a little piece of the sea,
from which some made saris, some made bandannas;
the rest was offered on trays to white cruise ships
taller than the post office; then the dogfights

began in the cabinets as to who had first sold
the archipelago for this chain store of islands.

Now a tree of grenades was his star-apple kingdom,
over fallow pastures his crows patrolled,
he felt his fist involuntarily tighten
into a talon that was strangling five doves,
the mountains loomed leaden under martial law,
the suburban gardens flowered with white paranoia
next to the bougainvilleas of astonishing April;
the rumours were a rain that would not fall:
that enemy intelligence had alerted the roaches'
quivering antennae, that bats flew like couriers,
transmitting secrets between the embassies;
over dials in the war rooms, the agents waited
for a rifle crack from Havana; down shuttered avenues
roared a phalanx of Yamahas. They left
a hole in the sky that closed on silence.

He didn't hear the roar of the motorcycles
diminish in circles like those of the water mill
in a far childhood; he was drowned in sleep;
he slept, without dreaming, the sleep after love
in the mineral oblivion of night
whose flesh smells of cocoa, whose teeth are white
as coconut meat, whose breath smells of ginger,
whose braids are scented like sweet-potato vines
in furrows still pungent with the sun.
He slept the sleep that wipes out history,
he slept like the islands on the breast of the sea,
like a child again in her star-apple kingdom.

Tomorrow the sea would gleam like nails
under a zinc sky where the barren frangipani
was hammered, a horizon without liners;
tomorrow the heavy caravels of clouds would wreck
and dissolve in their own foam on the reefs
of the mountains, tomorrow a donkey's yawn
would saw the sky in half, and at dawn
would come the noise of a government groaning uphill.
But now she held him, as she holds us all,
her history-orphaned islands, she to whom
we came late as our muse, our mother,
who suckled the islands, who, when she grows old
with her breasts wrinkled like eggplants,
is the head-tie mother, the bleached-sheets-on-the-river-rocks
    mother,
the gospel mother, the t'ank-you-parson mother
who turns into mahogany, the lignum-vitae mother,
her sons like thorns,
her daughters dry gullies that give birth to stones,
who was, in our childhood, the housemaid and the cook,
the young grand' who polished the plaster figure
of Clio, Muse of history, in her seashell grotto
in the Great House parlour, Anadyomene washed
in the deep Atlantic heave of her housemaid's hymn.

In the indigo dawn the palms unclenched their fists,
his eyes opened the flowers, and he lay as still
as the waterless mill wheel. The sun's fuse caught,
it hissed on the edge of the skyline, and day exploded
its remorseless avalanche of dray carts and curses,
the roaring oven of Kingston, its sky as fierce
as the tin box of a patties cart. Down the docks

between the Levantine smells of the warehouses
nosed the sea-wind with its odour of a dog's damp fur.
He lathered in anger and refreshed his love.
He was lathered like a horse, but the instant
the shower crowned him and he closed his eyes,
he was a bride under lace, remarrying his country,
a child drawn by the roars of the mill wheel's electorate,
those vows reaffirmed; he dressed, went down to breakfast,
and sitting again at the mahogany surface
of the breakfast table, its dark hide as polished
as the sheen of mares, saw his father's face
and his own face blent there, and looked out
to the drying garden and its seeping pond.

What was the Caribbean? A green pond mantling
behind the Great House columns of Whitehall,
behind the Greek façades of Washington,
with bloated frogs squatting on lily pads
like islands, islands that coupled as sadly as turtles
engendering islets, as the turtle of Cuba
mounting Jamaica engendered the Caymans, as, behind
the hammerhead turtle of Haiti–San Domingo
trailed the little turtles from Tortuga to Tobago;
he followed the bobbing trek of the turtles
leaving America for the open Atlantic,
felt his own flesh loaded like the pregnant beaches
with their moon-guarded eggs—they yearned for Africa,
they were lemmings drawn by magnetic memory
to an older death, to broader beaches
where the coughing of lions was dumbed by breakers.
Yes, he could understand their natural direction
but they would drown, sea-eagles circling them,

and the languor of frigates that do not beat wings,
and he closed his eyes, and felt his jaw drop
again with the weight of that silent scream.

He cried out at the turtles as one screams at children
with the anger of love, it was the same scream
which, in his childhood, had reversed an epoch
that had bent back the leaves of his star-apple kingdom,
made streams race uphill, pulled the water wheel backwards
like the wheels in a film, and at that outcry,
from the raw ropes and tendons of his throat,
the sea-buzzards receded and receded into specks,
and the osprey vanished.
                        On the knee-hollowed steps
of the crusted cathedral, there was a woman in black,
the black of moonless nights, within whose eyes
shone seas in starlight like the glint of knives
(the one who had whispered to the keyhole of his ear),
washing the steps, and she heard it first.
She was one of a flowing black river of women
who bore elliptical basins to the feet of paupers
on the Day of Thorns, who bore milk pails to cows
in a pastoral sunrise, who bore baskets on their heads
down the haemophilic red hills of Haiti,
now with the squeezed rag dripping from her hard hands
the way that vinegar once dropped from a sponge,
but she heard as a dog hears, as all the underdogs
of the world hear, the pitched shriek of silence.
Star-apples rained to the ground in that silence,
the silence was the green of cities undersea,
and the silence lasted for half an hour
in that single second, a seashell silence, resounding

with silence, and the men with barbed-wire beards saw
in that creak of light that was made between
the noises of the world that was equally divided
between rich and poor, between North and South,
between white and black, between two Americas,
the fields of silent Zion in Parish Trelawny,
in Parish St. David, in Parish St. Andrew,
leaves dancing like children without any sound,
in the valley of Tryall, and the white, silent roar
of the old water wheel in the star-apple kingdom;
and the woman's face, had a smile been decipherable
in that map of parchment so rivered with wrinkles,
would have worn the same smile with which he now
cracked the day open and began his egg.

From

THE FORTUNATE

TRAVELLER

[1981]

Old New England

Black clippers, tarred with whales' blood, fold their sails
entering New Bedford, New London, New Haven.
A white church spire whistles into space
like a swordfish, a rocket pierces heaven
as the thawed springs in icy chevrons race
down hillsides and Old Glories flail
the crosses of green farm boys back from 'Nam.
Seasons are measured still by the same
span of the veined leaf and the veined body
whenever the spring wind startles an uproar
of marching oaks with memories of a war
that peeled whole counties from the calendar.

The hillside is still wounded by the spire
of the white meetinghouse, the Indian trail
trickles down it like the brown blood of the whale
in rowanberries bubbling like the spoor
on logs burnt black as Bibles by hellfire.
The war whoop is coiled tight in the white owl,
stone-feathered icon of the Indian soul,
and railway lines are arrowing to the far
mountainwide absence of the Iroquois.
Spring lances wood and wound, and a spring runs
down tilted birch floors with their splintered suns

of beads and mirrors—broken promises
that helped make this Republic what it is.

The crest of our conviction grows as loud
as the spring oaks, rooted and reassured
that God is meek but keeps a whistling sword;
His harpoon is the white lance of the church,
His wandering mind a trail folded in birch,
His rage the vats that boiled the melted beast
when the black clippers brought (knotting each shroud
round the crosstrees) our sons home from the East.

Upstate

A knife blade of cold air keeps prying
the bus window open. The spring country
won't be shut out. The door to the john
keeps banging. There're a few of us:
a stale-drunk or stoned woman in torn jeans,
a Spanish-American salesman, and, ahead,
a black woman folded in an overcoat.
Emptiness makes a companionable aura
through the upstate villages—repetitive,
but crucial in their little differences
of fields, wide yards with washing, old machinery—where
    people live
with the highway's patience and flat certainty.

Sometimes I feel sometimes
the Muse is leaving, the Muse is leaving America.
Her tired face is tired of iron fields,
its hollows sing the mines of Appalachia,
she is a chalk-thin miner's wife with knobbled elbows,
her neck tendons taut as banjo strings,
she who was once a freckled palomino with a girl's mane
galloping blue pastures plinkety-plunkety,
staring down at a tree-stunned summer lake,
when all the corny calendars were true.

The departure comes over me in smoke
from the far factories.

But were the willows lyres, the fanned-out pollard willows
with clear translation of water into song,
were the starlings as heartbroken as nightingales,
whose sorrow piles the looming thunderhead
over the Catskills, what would be their theme?
The spring hills are sun-freckled, the chaste white barns flash
through screening trees the vigour of her dream,
like a white plank bridge over a quarrelling brook.
Clear images! Direct as your daughters
in the way their clear look returns your stare,
unarguable and fatal—
no, it is more sensual.
I am falling in love with America.

I must put the cold small pebbles from the spring
upon my tongue to learn her language,
to talk like birch or aspen confidently.
I will knock at the widowed door
of one of these villages
where she will admit me like a broad meadow,
like a blue space between mountains,
and holding her arms at the broken elbows
brush the dank hair from a forehead
as warm as bread or as a homecoming.

Piano Practice

[*for Mark Strand*]

April, in another fortnight, metropolitan April.
A drizzle glazes the museum's entrance,
like their eyes when they leave you, equivocating spring!
The sun dries the avenue's pumice façade
delicately as a girl tamps tissue on her cheek;
the asphalt shines like a silk hat,
the fountains trot like percherons round the Met,
clip, clop, clip, clop in Belle Epoque Manhattan,
as gutters part their lips to the spring rain—
down avenues hazy as Impressionist clichés,
their gargoyle cornices,
their concrete flowers on chipped pediments,
their subway stops in Byzantine mosaic—
the soul sneezes and one tries to compile
the collage of a closing century,
the epistolary pathos, the old Laforguean ache.

Deserted plazas swept by gusts of remorse,
rain-polished cobbles where a curtained carriage
trotted around a corner of Europe for the last time,
as the canals folded like concertinas.
Now fever reddens the trouble spots of the globe,
rain drizzles on the white iron chairs in the gardens.

Today is Thursday, Vallejo is dying,
but come, girl, get your raincoat, let's look for life
in some café behind tear-streaked windows,
perhaps the *fin de siècle* isn't really finished,
maybe there's a piano playing it somewhere,
as the bulbs burn through the heart of the afternoon
in the season of tulips and the pale assassin.
I called the Muse, she pleaded a headache,
but maybe she was just shy at being seen
with someone who has only one climate,
so I passed the flowers in stone, the sylvan pediments,
alone. It wasn't I who shot the archduke,
I excuse myself of all crimes of that ilk,
muttering the subway's obscene graffiti;
I could offer her nothing but the predictable
pale head-scarf of the twilight's lurid silk.

Well, goodbye, then, I'm sorry I've never gone
to the great city that gave Vallejo fever.
Maybe the Seine outshines the East River,
maybe, but near the Metropolitan
a steel tenor pan
dazzlingly practices something from old Vienna,
the scales skittering like minnows across the sea.

North and South

Now, at the rising of Venus—the steady star
that survives translation, if one can call this lamp
the planet that pierces us over indigo islands—
despite the critical sand flies, I accept my function
as a colonial upstart at the end of an empire,
a single, circling, homeless satellite.
I can listen to its guttural death rattle in the shoal
of the legions' withdrawing roar, from the raj,
from the Reich, and see the full moon again
like a white flag rising over Fort Charlotte,
and sunset slowly collapsing like the flag.

It's good that everything's gone, except their language,
which is everything. And it may be a childish revenge
at the presumption of empires to hear the worm
gnawing their solemn columns into coral,
to snorkel over Atlantis, to see, through a mask,
Sidon up to its windows in sand, Tyre, Alexandria,
with their wavering seaweed spires through a glass-bottom boat,
and to buy porous fragments of the Parthenon
from a fisherman in Tobago, but the fear exists,
Delenda est Carthago on the rose horizon,

and the side streets of Manhattan are sown with salt,
as those in the North all wait for that white glare

of the white rose of inferno, all the world's capitals.
Here, in Manhattan, I lead a tight life
and a cold one, my soles stiffen with ice
even through woollen socks; in the fenced back yard,
trees with clenched teeth endure the wind of February,
and I have some friends under its iron ground.
Even when spring comes with its rain of nails,
with its soiled ice oozing into black puddles,
the world will be one season older but no wiser.

Fragments of paper swirl round the bronze general
of Sheridan Square, syllables of Nordic tongues
(as an obeah priestess sprinkles flour on the doorstep
to ward off evil, so Carthage was sown with salt);
the flakes are falling like a common language
on my nose and lips, and rime forms on the mouth
of a shivering exile from his African province;
a blizzard of moths whirls around the extinguished lamp
of the Union general, sugary insects crunched underfoot.

You move along dark afternoons where death
entered a taxi and sat next to a friend,
or passed another a razor, or whispered "Pardon"
in a check-clothed restaurant behind her cough—
I am thinking of an exile farther than any country.
And, in this heart of darkness, I cannot believe
they are now talking over palings by the doddering
banana fences, or that seas can be warm.

How far I am from those cacophonous seaports
built round the single exclamation of one statute
of Victoria Regina! There vultures shift on the roof

of the red iron market, whose patois
is brittle as slate, a grey stone flecked with quartz.
I prefer the salt freshness of that ignorance,
as language crusts and blackens on the pots
of this cooked culture, coming from a raw one;
and these days in bookstores I stand paralyzed

by the rows of shelves along whose wooden branches
the free-verse nightingales are trilling "Read me! Read me!"
in various metres of asthmatic pain;
or I shiver before the bellowing behemoths
with the snow still falling in white words on Eighth Street,
those burly minds that barrelled through contradictions
like a boar through bracken, or an old tarpon
bristling with broken hooks, or an old stag
spanielled by critics to a crag at twilight,

the exclamation of its antlers like a hat rack
on which they hang their theses. I am tired of words,
and literature is an old couch stuffed with fleas,
of culture stuffed in the taxidermist's hides.
I think of Europe as a gutter of autumn leaves
choked like the thoughts in an old woman's throat.
But she was home to some consul in snow-white ducks
doing out his service in the African provinces,
who wrote letters like this one home and feared malaria
as I mistrust the dark snow, who saw the lances of rain

marching like a Roman legion over the fens.
So, once again, when life has turned into exile,
and nothing consoles, not books, work, music, or a woman,
and I am tired of trampling the brown grass,

whose name I don't know, down an alley of stone,
and I must turn back to the road, its winter traffic,
and others sure in the dark of their direction,
I lie under a blanket on a cold couch,
feeling the flu in my bones like a lantern.

Under the blue sky of winter in Virginia
the brick chimneys flute white smoke through skeletal lindens,
as a spaniel churns up a pyre of blood-rusted leaves;
there is no memorial here to their Treblinka—
as a van delivers from the ovens loaves
as warm as flesh, its brakes jaggedly screech
like the square wheel of a swastika. The mania
of history veils even the clearest air,
the sickly sweet taste of ash, of something burning.

And when one encounters the slow coil of an accent,
reflexes step aside as if for a snake,
with the paranoid anxiety of the victim.
The ghosts of white-robed horsemen float through the trees,
the galloping hysterical abhorrence of my race—
like any child of the Diaspora, I remember this
even as the flakes whiten Sheridan's shoulders,
and I remember once looking at my aunt's face,
the wintry blue eyes, the rusty hair, and thinking

maybe we are part Jewish, and felt a vein
run through this earth and clench itself like a fist
around an ancient root, and wanted the privilege
to be yet another of the races they fear and hate
instead of one of the haters and the afraid.
Above the spiny woods, dun grass, skeletal trees,

the chimney serenely fluting something from Schubert—
like the wraith of smoke that comes from someone burning—
veins the air with an outcry that I cannot help.

The winter branches are mined with buds,
the fields of March will detonate the crocus,
the olive battalions of the summer woods
will shout orders back to the wind. To the soldier's mind
the season's passage round the pole is martial,
the massacres of autumn sheeted in snow, as
winter turns white as a veterans hospital.
Something quivers in the blood beyond control—
something deeper than our transient fevers.

But in Virginia's woods there is also an old man
dressed like a tramp in an old Union greatcoat,
walking to the music of rustling leaves, and when
I collect my change from a small-town pharmacy,
the cashier's fingertips still wince from my hand
as if it would singe hers—well, yes, *je suis un singe*,
I am one of that tribe of frenetic or melancholy
primates who made your music for many more moons
than all the silver quarters in the till.

Beachhead

[*for Tony Hecht*]

I come up to a break
on the beach where a channel
of the river is pushed back
by the ancestral quarrel

of fresh water with salt.
Under it: scalloped sand.
Not caring who's at fault,
I turn and cross inland.

A sepia lagoon
bobbing with coconuts—
helmets from the platoon
of some Marine unit—

whose channel links those years
of boyhood photographs
in *Life* or *Collier's*
to dim Pacific surf.

Sandpipers burst like white
notes from a ceremonial band,

circle, then, on wet sand,
discuss their cancelled flight.

The beach is hot, the fronds
of yellow dwarf palms rust,
the clouds are close as friends,
the sea has not learned rest,

exploding, but not in,
thank heaven, that rhetoric
all wars must be fought in,
I break a brittle stick

pointlessly and walk on,
holding the stick, until
it hefts like a weapon.
There is nothing to kill.

Guadalcanal and Guam—
they must now look like this
abandoned Navy base
camouflaged in gold palm.

Divisions, dates, and armour
marked here are not enough.
The surf, a plasterer,
smooths a fresh cenotaph.

I hurl the stick and brush
right hand against left hand.
Snipers prowl through the bush
of my dry hair. I stand,

not breathing, till they pass,
and the new world feels sure:
sand and sand-whitened grass,
then a jet's signature.

Map of the New World

I *Archipelagoes*

At the end of this sentence, rain will begin.
At the rain's edge, a sail.

Slowly the sail will lose sight of islands;
into a mist will go the belief in harbours
of an entire race.

The ten-years war is finished.
Helen's hair, a grey cloud.
Troy, a white ashpit
by the drizzling sea.

The drizzle tightens like the strings of a harp.
A man with clouded eyes picks up the rain
and plucks the first line of the *Odyssey*.

From This Far

[*for Robert Giroux*]

I

The white almonds of a statue stare
at almond branches wrestling off their shade
like a girl from her dress—a gesture rarely made
by abstract stone.
              A Greek tanker passes
through the net of branches
to the drag of tractors quarrying a cliff—
in its hold, a cargo of marble heads;
from Orpheus to Onassis,
the sea has flown one flag:
white-barred waves on unalterable blue.

The sky's window rattles
at gears raked into reverse;
but no stone head rolls in the ochre dust,
in the soil of our islands no gods are buried.
They were shipped to us, Seferis,
dead on arrival.

II

Dawn buckles on the helmet
of rayed Agamemnon.
A net is flung over the shallows;
ocean divides: a bronze door.
In the wash the trunks of warriors
roll and recede.
Great lines, Seferis, have heaved them this far.

At dusk, the man-god bleeds
face down in the veins of the sea.

The blue night hums with bees.
Every hour bores a hole in the hive
of the labyrinth, at whose end
the obscene miscegenation lowers its lyre-curved horns,
and whether it is for dead stones, or the god of thorns,
we stagger the arena with leaking eyes.

The almonds hoard their shadows
as we do the shades of friends.
When a bronze leaf glints, I hear again
the torn throat in the torn shade,
then my eyes harden in a stone head.

I see them in a colonnade
of concrete wharf-piles
where a gull settles.
I hear them groaning with the tractors.
I am eating an ice cream on a hot esplanade,
in a barred blue-and-white vest,

in the brittle shade of a sea grape,
in the iodine reek of shallows,
watching the empty blue port
frothing with yachts,
when a leafy wall
tosses the shadow of a pawing bull.
The Ferryboat passes,
and the gull screeches its message,
opening its wings like a letter,
and the screech grows into a whirlwind
of shawled and ragged crows in a stone field.

III

It is during this, Seferis,
that a girl wrestling off her dress
folds with the wave like a dolphin,
that surf hides the sobbing of women,
that, in the thudding of tractors,
I hear the wooden clogs
behind the hills' arena,
and the dry retching of the hunting dogs.

Over something—carrion, the sun's wave-buried king—
vultures with ragged shawls keep circling;
I see the harpist with his eyes like clouds,
I remember you holding a heavy marble head;
I see the other who invited the barbarians
into the whitewashed streets.

I stayed with my own. I starved my hand of names,
no tan fauns leapt over my wrist,

I'll never see Piraeus repeat her white name in water,
but whether my eyes will be white seeds in a bust,
or, likelier, the salt fruit of worms,
they are sockets whose hollows boast
those flashes of inward life,
from the head's thunder-lit storms.

Europa

The full moon is so fierce that I can count the
coconuts' cross-hatched shade on bungalows,
their white walls raging with insomnia.
The stars leak drop by drop on the tin plates
of the sea almonds, and the jeering clouds
are luminously rumpled as the sheets.
The surf, insatiably promiscuous,
groans through the walls; I feel my mind
whiten to moonlight, altering that form
which daylight unambiguously designed,
from a tree to a girl's body bent in foam;
then, treading close, the black hump of a hill,
its nostrils softly snorting, nearing the
naked girl splashing her breasts with silver.
Both would have kept their proper distance still,
if the chaste moon hadn't swiftly drawn the drapes
of a dark cloud, coupling their shapes.

She teases with those flashes, yes, but once
you yield to human horniness, you see
through all that moonshine what they really were,
those gods as seed-bulls, gods as rutting swans—
an overheated farmhand's literature.
Who ever saw her pale arms hook his horns,
her thighs clamped tight in their deep-plunging ride,

watched, in the hiss of the exhausted foam,
her white flesh constellate to phosphorus
as in salt darkness beast and woman come?
Nothing is there, just as it always was,
but the foam's wedge to the horizon-light,
then, wire-thin, the studded armature,
like drops still quivering on his matted hide,
the hooves and horn-points anagrammed in stars.

The Man Who Loved Islands

A Two-Page Outline

A man is leaning on a cold iron rail
watching an islet from an island and so on,
say, Charlotte Amalie facing St. John,
which begins the concept of infinity
uninterrupted by any mortal sail,
only the thin ghost of a tanker drawing the horizon
behind it with the silvery slick of a snail,
and that's the first shot of this forthcoming film
starring James Coburn and his tanned, leathery, frail
resilience and his now whitening hair,
and his white, vicious grin. Now, we were where?
On this island, one of the Virgins, the prota-
gonist established. Now comes the second shot,
and chaos of artifice still called the plot,
which has to get the hero off somewhere
else, 'cause there's no kick in contemplation
of silvery light upon wind-worried water
between here and the islet of St. John,
and how they are linked like any silver chain
glinting against the hero's leather chest,
sold in the free gift ports, like noon-bright water.
The hero's momentary rest on the high rail
can be a good beginning. To start with rest
is good—the tanker can come later.

But we can't call it "The Man Who Loved Islands"
any more than some Zen–Karate film
would draw them with "The Hero Who Loves Water."
No soap. There must be something with diamonds,
emeralds, emeralds the color of the shallows there,
or sapphires, like blue unambiguous air,
sapphires for Sophia, but we'll come to that.
Coburn looks great with or without a hat,
and there must be some minimum of slaughter
that brings in rubies, but you cannot hover
over that first shot like a painting. Action
is all of art, the thoughtless pace
of lying with style, so that when it's over,
that first great shot of Coburn's leathery face,
crinkled like the water which he contemplates,
could be superfluous, in the first place,
since that tired artifice called history,
which in its motion is as false as fiction,
requires an outline, a summary. I can think of none,
quite honestly. I'm no photographer; this
could be a movie. I mean things are moving,
the water for example, the light on the man's hair
that has gone white, even those crescent sands
are just as moving as his love of islands;
the tanker that seems still is moving, even
the clouds like galleons anchored in heaven,
and what is moving most of all of course
is the violent man lulled into this inaction
by the wide sea. Let's hold it on the sea
as we establish their ancient interaction,
a hint of the Homeric, a little poetry
before the whole mess hits the bloody fan.

All these islands that you love, I guaran-
tee we'll work them in as background, with
generous establishing shots from Jim's car and
even a few harbours and villages, *if*
we blow the tanker up and get the flames
blazing with oil, and Sophia, if she's free,
daintily smudged, with her slip daintily torn,
is climbing down this rope ladder, and we shoot up
from Coburn's P.O.V.—he's got the gems—
that's where we throw in Charlotte Amalie
and the waterfront bars, and this Danish alley
with the heavies chasing, and we can keep all the
business of Jim on the rail; that lyric stuff
goes with the credits if you insist on keeping it tend-
er; I can see it, but things must get rough
pretty damn fast, or else you lose them, pally,
or, tell you what, let's save it for THE END.

Hurucan

I

Once branching light startles the hair of the coconuts,
and on the villas' asphalt roofs, rain
resonates like pebbles in a pan,
and only the skirts of surf
waltz round the abandoned bandstand,
and we hear the telephone cables
hallooing like fingers tapped over an Indian's mouth,
once the zinc roofs begin wrenching their nails
like freight uncrated with a crowbar,
we remember you as the possible
deity of the whistling marsh-canes,
we doubt that you were ever slain
by the steel Castilian lances
of a thousand horizons,
deity of the yellow-skinned ones
who thatched your temple with plantains.

When the power station's blackout
grows frightening as amnesia,
and the luxury resorts
revert to the spear-tips of candles,
and the swimming pools in their marsh-light
multiply with hysterical lilies
like the beaks of fledglings uttering your name,

when lightning fizzles out
in the wireless, we can see and hear
the streaming black locks of clouds,
flesh the gamboge of lightning,
and the epicanthic, almond-shaped eye
of the whirling cyclops,
runner through the cloud-smoke,
our ocean's marathon strider,
the only survivor
of their massacred deities

whose temples change
like the clouds over Yucatán
in the copper twilight over Ecuador,
runner who can grip the mares' tails
of galloping cirrus,
vaulting the dead conquistadors of the helmeted palms.
You'd never reply
to the name of the northern messenger
whose zigzagging trident
pitchforks the oaks like straw,
nor to the thunderous tambours
of Shango; you rage
till we get your name right,
till the surf and the bent palms dance
to your tune, even if, at your entrance,
clouds plod the horizon like caparisoned camels,
and the wind begins to unwhirl
like a burnoose; you abhor
all other parallels
but our own,
Hurucan.

You scream like a man whose wife is dead,
like a god who has lost his race,
you yank the electric wires with wet hands.

Then we think of a different name
than the cute ones christened by radar,
in the sludge that sways
next day by the greased pierheads
where a rowboat still rocks in fear,
and Florida now flares to your flashbulb
and the map of Texas rattles,
and we lie awake in the dark
by the dripping stelae of candles,
our heads gigantified on the walls,
and think you, still running
with tendons feathered with lightning,
water-worrier, whom the chained trees
strain to follow,
havoc, reminder, ancestor,
and, when morning enters, pale
as an insurance broker,
god.

   II

The sea almond's dress
is drenched in the morning,
the leaves drip on their clotheslines
like wax drops from candles,
the pent waves circle their fences
like witless sheep.
A freighter is parked

on the coastal road to the airport,
and the birds won't be back
for some time. The chairs
around the bandstand are heaped up
like the morning after your dance,
and the worms we have buried underground
spark and stutter again. Roofs
are scattered all over the hillsides
like cards dropped during a shoot-out,
and the sea starts the pompous thunder
of a military funeral
as spray shoots up round the kiosk
where the Police Band played.

We return the pieces of fear
to their proper place,
the shelf at the back of the mind—
the artifacts, the Carib arrowheads,
the pin-pierced amulets,
and that force whose weather vanes
are the slow-spinning frigates.
Your name fades again in the grounded
flights; there in dark hangars
the mineral patience of cattle—
a cold sweat slides down the silver
hides of the empty planes.

Jean Rhys

In their faint photographs
mottled with chemicals,
like the left hand of some spinster aunt,
they have drifted to the edge
of verandahs in Whistlerian
white, their jungle turned tea-brown—
even its spiked palms—
their features pale,
to be pencilled in:
bone-collared gentlemen
with spiked moustaches
and their wives embayed in the wickerwork
armchairs, all looking coloured
from the distance of a century
beginning to groan sideways from the axe stroke!

Their bay horses blacken
like spaniels, the front lawn a beige
carpet, brown moonlight and a moon
so sallow, so pharmaceutical
that her face is a feverish child's,
some malarial angel
whose grave still cowers
under a fury of bush,

a mania of wild yams
wrangling to hide her from ancestral churchyards.

And the sigh of that child
is white as an orchid
on a crusted log
in the bush of Dominica,
a V of Chinese white
meant for the beat of a seagull
over a sepia souvenir of Cornwall,
as the white hush between two sentences.

Sundays! Their furnace
of boredom after church.
A maiden aunt canoes through lilies of clouds
in a Carib hammock, to a hymn's metronome,
and the child on the varnished, lion-footed couch
sees the hills dip and straighten with each lurch.
The green-leaved uproar of the century
turns dim as the Atlantic, a rumourous haze
behind the lime trees, breakers
advancing in decorous, pleated lace;
the cement grindstone of the afternoon
turns slowly, sharpening her senses,
the bay below is green as calalu, stewing Sargasso.

In that fierce hush
between Dominican mountains
the child expects a sound
from a butterfly clipping itself to a bush
like a gold earring to a black maid's ear—

one who goes down to the village, visiting,
whose pink dress wilts like a flower between the limes.

There are logs
wrinkled like the hand of an old woman
who wrote with a fine courtesy to that world
when grace was common as malaria,
when the gas lanterns' hiss on the verandah
drew the aunts out like moths
doomed to be pressed in a book, to fall
into the brown oblivion of an album,
embroiderers of silence
for whom the arches of the Thames,
Parliament's needles,
and the petit-point reflections of London Bridge
fade on the hammock cushions from the sun,
where one night
a child stares at the windless candle flame
from the corner of a lion-footed couch
at the erect white light,
her right hand married to *Jane Eyre*,
foreseeing that her own white wedding dress
will be white paper.

The Liberator

In a blue bar at the crossroads, before you turn
into Valencia or Grande, Castilian bequests,
in back of that bar, cool and dark as prison,
where a sunbeam dances through brown rum-bottles
like a firefly through a thicket of cocoa,
like an army torch looking for a guerrilla,
the guerrilla with the gouged Spanish face named
Sonora again climbs the track through wild bananas,
sweat glued to his face like a hot cloth
under the barber's hand. The jungle is steam.
He would like to plunge his hands in those clouds
on the next range. From Grande to Valencia
the blue-green plain below breaks through the leaves.
"Adios, then," said Estenzia. He went downhill.
And the army find him. The world keep the same.
The others get tired of just eating mango,
and fig, carefully fried. They dream of beds.
They bawl for their mudder and their children haunt them.
They dream of mattresses, even those in prison.
For half an hour I, Sonora, sit in the sun
till my face turn copper. And for that half hour,
the spikes of wild palm was Pizarro helmet,
and for that half hour, señor, I ain't look at them.

A fly, big like a bee, dance on my rifle barrel
like he know who was holding it already dead.
I turn: Manuel. Frederico. Marcos. The people
of Hernando Cortes, of Pizarro, El Cid—
men who had flung the pennoned spear flying
to the oaken door, the heart of authority.
We was going so good. But then, they get tired.
For a straight double-rum at a wet zinc table
in a blue bar at the crossroads, before you turn
into Valencia or Grande, Castilian bequests,
Sonora, the socialist, on any given Sunday
will narrate this adventure, which, inevitably,
a loss of heredity needs to create.

The Spoiler's Return

[*for Earl Lovelace*]

I sit high on this bridge in Laventille,
watching that city where I left no will
but my own conscience and rum-eaten wit,
and limers passing see me where I sit,
ghost in brown gabardine, bones in a sack,
and bawl: "Ay, Spoiler, boy! When you come back?"
And those who bold don't feel they out of place
to peel my limeskin back, and see a face
with eyes as cold as a dead macajuel,
and if they still can talk, I answer: "Hell."
I have a room there where I keep a crown,
and Satan send me to check out this town.
Down there, that Hot Boy have a stereo
where, whole day, he does blast my caiso;
I beg him two weeks' leave and he send me
back up, not as no bedbug or no flea,
but in this limeskin hat and floccy suit,
to sing what I did always sing: the truth.
Tell Desperadoes when you reach the hill,
I decompose, but I composing still:

I going to bite them young ladies, partner,
like a hot dog or a hamburger

and if you thin, don't be in a fright
is only big fat women I going to bite.

The shark, racing the shadow of the shark
across clear coral rocks, does make them dark—
that is my premonition of the scene
of what passing over this Caribbean.
Is crab climbing crab-back, in a crab-quarrel,
and going round and round in the same barrel,
is sharks with shirt-jacs, sharks with well-pressed fins,
ripping we small-fry off with razor grins;
nothing ain't change but colour and attire,
so back me up, Old Brigade of Satire,
back me up, Martial, Juvenal, and Pope
(to hang theirself I giving plenty rope),
join Spoiler' chorus, sing the song with me,
Lord Rochester, who praised the nimble flea:

Were I, who to my cost already am
One of those strange, prodigious creatures, Man,
A spirit free, to choose for my own share,
What case of flesh and blood I pleased to wear,
I hope when I die, after burial,
To come back as an insect or animal.

I see these islands and I feel to bawl,
"area of darkness" with V. S. Nightfall.

Lock off your tears, you casting pearls of grief
on a duck's back, a waxen dasheen leaf,
the slime crab's carapace is waterproof
and those with hearing aids turn off the truth,

and their dark glasses let you criticize
your own presumptuous image in their eyes.
Behind dark glasses is just hollow skull,
and black still poor, though black is beautiful.
So, crown and mitre me Bedbug the First—
the gift of mockery with which I'm cursed
is just a insect biting Fame behind,
a vermin swimming in a glass of wine,
that, dipped out with a finger, bound to bite
its saving host, ungrateful parasite,
whose sting, between the cleft arse and its seat,
reminds Authority man is just meat,
a moralist as mordant as the louse
that the good husband brings from the whorehouse,
the flea whose itch to make all Power wince,
will crash a fête, even at his life's expense,
and these pile up in lime pits by the heap,
daily, that our deliverers may sleep.
All those who promise free and just debate,
then blow up radicals to save the state,
who allow, in democracy's defence,
a parliament of spiked heads on a fence,
all you go bawl out, "Spoils, things ain't so bad."
This ain't the Dark Age, is just Trinidad,
is human nature, Spoiler, after all,
it ain't big genocide, is just bohbohl;
safe and conservative, 'fraid to take side,
they say that Rodney commit suicide,
is the same voices that, in the slave ship,
smile at their brothers, "Boy, is just the whip,"
I free and easy, you see me have chain?
A little censorship can't cause no pain,

a little graft can't rot the human mind,
what sweet in goat-mouth sour in his behind.
So I sing with Attila, I sing with Commander,
what right in Guyana, right in Uganda.
The time could come, it can't be very long,
when they will jail calypso for picong,
for first comes television, then the press,
all in the name of Civic Righteousness;
it has been done before, all Power has
made the sky shit and maggots of the stars,
over these Romans lying on their backs,
the hookers swaying their enormous sacks,
until all language stinks, and the truth lies,
a mass for maggots and a fête for flies;
and, for a spineless thing, rumour can twist
into a style the local journalist—
as bland as a green coconut, his manner
routinely tart, his sources the Savannah
and all pretensions to a native art
reduced to giggles at the coconut cart,
where heads with reputations, in one slice,
are brought to earth, when they ain't eating nice;
and as for local Art, so it does go,
the audience have more talent than the show.

Is Carnival, straight Carnival that's all,
the beat is base, the melody bohbohl,
all Port of Spain is a twelve-thirty show,
some playing Kojak, some Fidel Castro,
some Rastamen, but, with or without locks,
to Spoiler is the same old khaki socks,
all Frederick Street stinking like a closed drain,

Hell is a city much like Port of Spain,
what the rain rots, the sun ripens some more,
all in due process and within the law,
as, like a sailor on a spending spree,
we blow our oil-bloated economy
on projects from here to eternity,
and Lord, the sunlit streets break Spoiler's heart,
to have natural gas and not to give a fart,
to see them line up, pitch-oil tin in hand:
each independent, oil-forsaken island,
like jeering at some scrunter with the blues,
while you lend him some need-a-half-sole shoes,
some begging bold as brass, some coming meeker,
but from Jamaica to poor Dominica
we make them know they begging, every loan
we send them is like blood squeezed out of stone,
and giving gives us back the right to laugh
that we couldn't see we own black people starve,
and, more we give, more we congratulate
we-self on our own self-sufficient state.
In all them project, all them Five-Year Plan,
what happen to the Brotherhood of Man?
Around the time I dead it wasn't so,
we sang the Commonwealth of caiso,
we was in chains, but chains made us unite,
now who have, good for them, and who blight, blight;
my bread is bitterness, my wine is gall,
my chorus is the same: "I want to fall."
Oh, wheel of industry, check out your cogs!
Between the knee-high trash and khaki dogs
Arnold's Phoenician trader reach this far,
selling you half-dead batteries for your car;

the children of Tagore, in funeral shroud,
curry favour and chicken from the crowd;
as for the Creoles, check their house, and look,
you bust your brain before you find a book,
when Spoiler see all this, ain't he must bawl,
"area of darkness," with V. S. Nightfall?
Corbeaux like cardinals line the La Basse
in ecumenical patience while you pass
the Beetham Highway—Guard corruption's stench,
you bald, black justices of the High Bench—
and beyond them the firelit mangrove swamps,
ibises practising for postage stamps,
Lord, let me take a taxi South again
and hear, drumming across Caroni Plain,
the tabla in the Indian half hour
when twilight fills the mud huts of the poor,
to hear the tattered flags of drying corn
rattle a sky from which all the gods gone,
their bleached flags of distress waving to me
from shacks, adrift like rafts on a green sea,
"Things ain't go change, they ain't go change at all,"
to my old chorus: "Lord, I want to bawl."
The poor still poor, whatever arse they catch.
Look south from Laventille, and you can watch
the torn brown patches of the Central Plain
slowly restitched by needles of the rain,
and the frayed earth, crisscrossed like old bagasse,
spring to a cushiony quilt of emerald grass,
and who does sew and sow and patch the land?
The Indian. And whose villages turn sand?
The fishermen doomed to stitching the huge net
of the torn foam from Point to La Fillette.

One thing with hell, at least it organize
in soaring circles, when any man dies
he must pass through them first, that is the style,
Jesus was down here for a little while,
cadaverous Dante, big-guts Rabelais,
all of them wave to Spoiler on their way.
Catch us in Satan tent, next carnival:
Lord Rochester, Quevedo, Juvenal,
Maestro, Martial, Pope, Dryden, Swift, Lord Byron,
the lords of irony, the Duke of Iron,
hotly contending for the monarchy
in couplets or the old re-minor key,
all those who gave earth's pompous carnival
fatigue, and groaned "O God, I feel to fall!"
all those whose anger for the poor on earth
made them weep with a laughter beyond mirth,
names wide as oceans when compared with mine
salted my songs, and gave me their high sign.
All you excuse me, Spoiler was in town;
you pass him straight, so now he gone back down.

The Hotel Normandie Pool

I

Around the cold pool in the metal light
of New Year's morning, I choose one of nine
cast-iron umbrellas set in iron tables
for work and coffee. The first cigarette
triggers the usual fusillade of coughs.
After a breeze the pool settles the weight
of its reflections on one line. Sunshine
lattices a blank wall with the shade of gables,
stirs the splayed shadows of the hills like moths.

Last night, framed in the binding of that window,
like the great chapter in some Russian novel
in which, during the war, the prince comes home
to watch the soundless waltzers dart and swivel,
like fishes in their lamplit aquarium,
I stood in my own gauze of swirling snow
and, through the parted hair of ribboned drapes,
felt, between gusts of music, the pool widen
between myself and those light-scissored shapes.

The dancers stiffened and, like fish, were frozen
in panes of ice blocked by the window frames;
one woman fanned, still fluttering on a pin,
as a dark fusillade of kettledrums

and a piercing cornet played "Auld Lang Syne"
while a battalion of drunk married men
reswore their vows. For this my fiftieth year,
I muttered to the ribbon-medalled water,
"Change me, my sign, to someone I can bear."

Now my pen's shadow, angled at the wrist
with the chrome stanchions at the pool's edge,
dims on its lines like birches in a mist
as a cloud fills my hand. A drop punctuates
the startled paper. The pool's iron umbrellas
ring with the drizzle. Sun hits the water.
The pool is blinding zinc. I shut my eyes,
and as I raise their lids I see each daughter
ride on the rayed shells of both irises.

The prayer is brief: That the transparent wrist
would not cloud surfaces with my own shadow,
and that this page's surface would unmist
after my breath as pools and mirrors do.
But all reflection gets no easier,
although the brown, dry needles of that palm
quiver to stasis and things resume their rhyme
in water, like the rubber ring that is a
red rubber ring inverted at the line's center.

Into that ring my younger daughter dived
yesterday, slithering like a young dolphin,
her rippling shadow hungering under her,
with nothing there to show how well she moved
but in my mind the veer of limb and fin.
Transparent absences! Love makes me look

through a clear ceiling into rooms of sand;
I ask the element that is my sign,
"Oh, let her lithe head through that surface break!"

Aquarian, I was married to water;
under that certain roof, I would lie still
next to my sister spirit, horizontal
below what stars derailed our parallel
from our far vow's undeviating course;
the next line rises as they enter it,
Peter, Anna, Elizabeth—Margaret
still sleeping with one arm around each daughter,
in the true shape of love, beyond divorce.

Time cuts down on the length man can endure
his own reflection. Entering a glass
I surface quickly now, prefer to breathe
the fetid and familiar atmosphere
of work and cigarettes. Only tyrants believe
their mirrors, or Narcissi, brooding on boards,
before they plunge into their images;
at fifty I have learnt that beyond words
is the disfiguring exile of divorce.

II

Across blue seamless silk, iron umbrellas
and a brown palm burn. A sandalled man comes out
and, in a robe of foam-frayed terry cloth,
with Roman graveness buries his room key,
then, mummy-oiling both forearms and face
with sunglasses still on, stands, fixing me,

and nods. Some petty businessman who tans
his pallor a negotiable bronze,
and the bright nod would have been commonplace

as he uncurled his shades above the pool's
reflecting rim—white towel, toga-slung,
foam hair repeated by the robe's frayed hem—
but, in the lines of his sun-dazzled squint,
a phrase was forming in that distant tongue
of which the mind keeps just a mineral glint,
the lovely Latin lost to all our schools:
"*Quis te misit, Magister?*" And its whisper went
through my cold body, veining it in stone.

On marble, concrete, or obsidian,
your visit, Master, magnifies the lines
of our small pool to that Ovidian
thunder of surf between the Baltic pines.
The light that swept Rome's squares and palaces,
washing her tangled fountains of green bronze
when you were one drop in a surf of faces—
a fleck of spittle from the she-wolf's tooth—
now splashes a palm's shadow at your foot.

Turn to us, Ovid. Our emerald sands
are stained with sewage from each tin-shacked Rome;
corruption, censorship, and arrogance
make exile seem a happier thought than home.
"Ah, for the calm proconsul with a voice
as just and level as this Roman pool,"
our house slaves sigh; the field slaves scream revenge;

one moves between the flatterer and the fool
yearning for the old bondage from both ends.

And I, whose ancestors were slave and Roman,
have seen both sides of the imperial foam,
heard palm and pine tree alternate applause
as the white breakers rose in galleries
to settle, whispering at the tilted palm
of the boy-god Augustus. My own face
held negro Neros, chalk Caligulas;
my own reflection slid along the glass
of faces foaming past triumphal cars.

Master, each idea has become suspicious
of its shadow. A lifelong friend whispers
in his own house as if it might arrest him;
markets no more applaud, as was their custom,
our camouflaged, booted militias
roaring past on camions, the sugar-apples
of grenades growing on their belts; ideas
with guns divide the islands; in dark squares
the poems gather like conspirators.

Then Ovid said, "When I was first exiled,
I missed my language as your tongue needs salt,
in every watery shape I saw my child,
no bench would tell my shadow 'Here's your place';
bridges, canals, willow-fanned waterways
turned from my parting gaze like an insult,
till, on a tablet smooth as the pool's skin,
I made reflections that, in many ways,
were even stronger than their origin.

"Tiled villas anchored in their foaming orchards,
parched terraces in a dust cloud of words,
among clod-fires, wolfskins, starving herds,
Tibullus' flute faded, sweetest of shepherds.
Through shaggy pines the beaks of needling birds
pricked me at Tomis to learn their tribal tongue,
so, since desire is stronger than its disease,
my pen's beak parted till we chirped one song
in the unequal shade of equal trees.

"Campaigns enlarged our frontiers like clouds,
but my own government was the bare boards
of a plank table swept by resinous pines
whose boughs kept skittering from Caesar's eye
with every yaw. There, hammering out lines
in that green forge to fit me for the horse,
I bent on a solitude so tyrannous
against the once seductive surf of crowds
that no wife softens it, or Caesar's envy.

"And where are those detractors now who said
that in and out of the imperial shade
I scuttled, showing to a frowning sun
the fickle dyes of the chameleon?
Romans"—he smiled—"will mock your slavish rhyme,
the slaves your love of Roman structures, when,
from Metamorphoses to Tristia,
art obeys its own order. Now it's time."
Tying his toga gently, he went in.

There, at the year's horizon, he had stood,
as if the pool's meridian were the line

that doubled the burden of his solitude
in either world; and, as one leaf fell,
his echo rippled: "Why here, of all places,
a small, suburban tropical hotel,
its pool pitched to a Mediterranean blue,
its palms rusting in their concrete oasis?
Because to make my image flatters you."

III

At dusk, the sky is loaded like watercolour paper
with an orange wash in which every edge frays—
a painting with no memory of the painter—
and what this pool recites is not a phrase
from an invisible, exiled laureate,
where there's no laurel, but the scant applause
of one dry, scraping palm tree as blue evening ignites its blossoms from one mango flower,
and something, not a leaf, falls like a leaf,

as swifts with needle-beaks dart, panicking over
the pool's cloud-closing light. For an envoi,
write what the wrinkled god repeats to the boy-
god: "May the last light of heaven pity us
for the hardening lie in the face that we did not tell."
Dusk. The trees blacken like the pool's umbrellas.
Dusk. Suspension of every image and its voice.
The mangoes pitch from their green dark like meteors.
The fruit bat swings on its branch, a tongueless bell.

Early Pompeian

[for Norline]

> Ere Babylon was dust,
> The Magus Zoroaster, my dead child,
> Met his own image walking in the garden,
> That apparition, sole of men, he saw.
> —SHELLEY

I

In the first years, when your hair
was parted severely in the Pompeian style,
you resembled those mosaics
whose round eyes
keep their immortal pinpoints, or were,
in laughing days, black olives on a saucer.

Then, one night, years later,
a flaring torch passed slowly down that wall
and lit them, and it was your turn.
Your girlhood was finished, your sorrows were robing
you with the readiness of woman.

The darkness placed a black shawl around your shoulders,
pointed to a colonnade of torches
like palm trees with their fronds on fire,

pointed out the cold flagstones to the sacrificial basin
where the priest stands with his birth-sword.
You nodded. You began to walk.

Voices stretched out their hands and you stepped from the wall.

Past the lowering eyes of rumors,
past the unblinking stares of the envious,
as, step by step, it faded
behind you, that portrait
with its plum-parted lips,
the skin of pomegranate,
the forehead's blank, unborn bewilderment.
Now you walked in those heel-hollowed steps
in which all of our mothers before us went.

And they led you, pale as the day-delivered moon,
through the fallen white columns of a hospital
to the volcanic bedrock of mud and screams and fire,

into the lava of the damned birth-blood,
the sacrificial gutters,
to where the eye of the stillborn star showed at the end of
    your road,
a dying star fighting the viruses
of furious constellations,
through the tangled veins, the vineyard of woman's labour,
to a black ditch under the corpuscles of stars,
where the shrunken grape would be born that would not call
    you mother.

In your noble, flickering gaze there was that which repeated
to the stone you carried
"The hardest times are the noblest, my dead child,"
and the torch passed its flame to your tongue,
your face bronzed in the drenches and fires of your finest sweat.

In their black sockets, the pebbles of your eyes
rattled like dice in the tin cup of the blind Fates.
On the black wings of your screams I watched vultures rise,
the laser-lances of pain splinter on the gods' breastplates.
Your nerve ends screamed like fifes,
your temples repeated a drum,
and your firelit head, in profile, passed other faces
as a funeral ship passes the torch-lit headlands
with its princely freight,
your black hair billowing like dishevelled smoke.

Your eyelids whitened like knuckles gripping
the incomprehensible, vague sills of pain.
The door creaked, groaning open, and in its draft, no, a whirlwind,
the lamp that was struggling with darkness was blown out
by the foul breeze off the amniotic sea.

II

By the black harbour,
the black schooners are tired
of going anywhere; the sea
is black and salt as the mind of a woman after labour.

Child, wherever you are,
I am still your father;

let your small, dead star
rock in my heart's black salt,
this sacrificial basin where I weep;
you passed from a sleep to a sleep
with no pilot, without a light.

Beautiful, black, and salt-warm is the starry night,
the smell off the sea is your mother,
as is this wind that moves in the leaves of the wharf under the
    pavement light.

I stare into black water by whose hulls
heaven is rocked like a cradle,
except, except for one extinguished star,
and I think of a hand that stretches out from her bedside
    for nothing,
and then is withdrawn, remembering where you are.

III

I will let the nights pass,
I shall allow the sun to rise,
I shall let it pass like a torch along a wall
on which there is fadingly set,
stone by fading stone,
the face of an astonished girl, her lips, her black hair parted
in the early Pompeian style.

And what can I write for her
but that when we are stoned with pain,
and we shake our heads wildly from side to side,
saying "no more," "no more again," to certain things,

no more faith, no more hope, only charity,
charity gives faith and hope much stronger wings.

    IV

As for you, little star,
my lost daughter, you are
bent in the shape forever
of a curled seed sailing the earth,
in the shape of one question, a comma
that knows before us whether death
is another birth.
                  I had no answer
to that tap-tapping under the dome
of the stomach's round coffin.
I could not guess whether you were calling
to be let in, or to be let go
when the door's groaning blaze
seared the grape-skin
frailty of your eyes crying
against our light, and all that is kin
to the light.
You had sailed without any light
your seven months on the amniotic sea.
You never saw your murderer,
your birth and death giver,
but I will see you everywhere,
I will see you in a boneless
sunbeam that strokes the texture
of things—my arm, the pulseless arm
of an armchair, an iron railing, the leaves
of a dusty plant by a closed door,

in the beams of my own eyes in a mirror.
The lives that we must go on with
are also yours. So I go on
down the apartment steps to the hot
streets of July the twenty-second, nineteen
hundred and eighty, in Trinidad,
amazed that trees are still green
around the Savannah, over the Queen's
Park benches, amazed that my feet can carry
the stone of the earth, the heavier stone of the head,
and I pass through shade where a curled
blossom falls from a black, forked branch
to the asphalt, soundlessly. No cry.
You knew neither this world nor the next,
and, as for us, whose hearts must never harden
against ourselves, who sit on a park bench
like any calm man in a public garden
watching the bright traffic,
we can only wonder why a seed should envy
our suffering, to flower, to suffer,
to die. Gloria, Perdita, I christen
you in the shade, on the bench,
with no hope of the resurrection.
Pardon. Pardon the pride I have taken
in a woman's agony.

Easter

Anna, my daughter,
you have a black dog
that noses your heel,
selfless as a shadow;
here is a fable
about a black dog:
On the last sunrise
the shadow dressed with Him,
it stretched itself also—
they were two big men
with one job to do.
But life had been lent to one
only for this life.
They strode in silence toward
uncontradicting night.
The rats at the Last Supper
shared crumbs with their shadows,
the shadow of the bread
was shared by the bread;
when the candles lowered,
the shadow felt larger,
so He ordered it to leave;
He said where He was going
it would not be needed,
for there there'd be either

radiance or nothing.
It stopped when He turned
and ordered it home,
then it resumed the scent;
it felt itself stretching
as the sun grew small
like the eyes of the soldiers
receding into holes
under the petrified
serpents on their helmets;
the narrowing pupils
glinted like nailheads,
so before He lay back
it crept between the wood
as if it were the pallet
they had always shared;
it crept between the wood
and the flesh nailed to the wood
and it rose like a black flag
as the crossbeam hoisted
itself and the eyes
closed very slowly
extinguishing the shadow—
everything was nothing.
Then the shadow slunk away,
crawling low on its belly,
and it left there knowing
that never again
would He ever need it;
it reentered the earth,
it didn't eat for three days,
it didn't go out,

then it peeped out carefully
like a mole from its hole,
like a wolf after winter,
like a surreptitious serpent,
looking for those forms
that could give back its shape;
then it ran out when the bells
began making wide rings
and rings of radiance;
it keeps nosing for His shape
and it finds it again, in
the white echo of a pigeon
with its wings extended
like a shirt on a clothesline,
like a white shirt on Monday
dripping from a clothesline,
like the greeting of a scarecrow
or a man yawning
at the end of a field.

Wales

[*for Ned Thomas*]

Those white flecks cropping the ridges of Snowdon
will thicken their fleece and come wintering down
through the gap between alliterative hills,
through the caesura that let in the Legions,
past the dark disfigured mouths of the chapels,
till a white silence comes to green-throated Wales.
Down rusty gorges, cold rustling gorse,
over rocks hard as consonants, and rain-vowelled shales
sang the shallow-buried axe, helmet, and baldric
before the wet asphalt sibilance of tires.
A plump raven, Plantagenet, unfurls its heraldic
caw over walls that held the cult of the horse.
In blackened cottages with their stony hatred
of industrial fires, a language is shared
like bread to the mouth, white flocks to dark byres.

The Fortunate Traveller

[*for Susan Sontag*]

And I heard a voice in the midst of the four beasts say,
A measure of wheat for a penny,
and three measures of barley for a penny;
and see thou hurt not the oil and the wine.
— REVELATION 6:6

I

It was in winter. Steeples, spires
congealed like holy candles. Rotting snow
flaked from Europe's ceiling. A compact man,
I crossed the canal in a grey overcoat,
on one lapel a crimson buttonhole
for the cold ecstasy of the assassin.
In the square coffin manacled to my wrist:
small countries pleaded through the mesh of graphs,
in treble-spaced, Xeroxed forms to the World Bank
on which I had scrawled the one word, MERCY;

       I sat on a cold bench
under some skeletal lindens.
Two other gentlemen, black skins gone grey
as their identical, belted overcoats,
crossed the white river.

They spoke the stilted French
of their dark river,
whose hooked worm, multiplying its pale sickle,
could thin the harvest of the winter streets.
"Then we can depend on you to get us those tractors?"
"I gave my word."
"May my country ask you why you are doing this, sir?"
Silence.
"You know if you betray us, you cannot hide?"
A tug. Smoke trailing its dark cry.

At the window in Haiti, I remember
a gecko pressed against the hotel glass,
with white palms, concentrating head.
With a child's hands. Mercy, monsieur. Mercy.
Famine sighs like a scythe
across the field of statistics and the desert
is a moving mouth. In the hold of this earth
10,000,000 shoreless souls are drifting.
Somalia: 765,000, their skeletons will go under the tidal sand.
"We'll meet you in Bristol to conclude the agreement?"
Steeples like tribal lances, through congealing fog
the cries of wounded church bells wrapped in cotton,
grey mist enfolding the conspirator
like a sealed envelope next to its heart.

No one will look up now to see the jet
fade like a weevil through a cloud of flour.
One flies first-class, one is so fortunate.
Like a telescope reversed, the traveller's eye
swiftly screws down the individual sorrow
to an oval nest of antic numerals,

and the iris, interlocking with this globe,
condenses it to zero, then a cloud.
Beetle-black taxi from Heathrow to my flat.
We are roaches,
riddling the state cabinets, entering the dark holes
of power, carapaced in topcoats,
scuttling around columns, signalling for taxis,
with frantic antennae, to other huddles with roaches;
we infect with optimism, and when
the cabinets crack, we are the first
to scuttle, radiating separately
back to Geneva, Bonn, Washington, London.

Under the dripping planes of Hampstead Heath,
I read her letter again, watching the drizzle
disfigure its pleading like mascara. Margo,
I cannot bear to watch the nations cry.
Then the phone: "We will pay you in Bristol."
Days in fetid bedclothes swallowing cold tea,
the phone stifled by the pillow. The telly
a blue storm with soundless snow.
I'd light the gas and see a tiger's tongue.
I was rehearsing the ecstasies of starvation
for what I had to do. *And have not charity*.

I found my pity, desperately researching
the origins of history, from reed-built communes
by sacred lakes, turning with the first sprocketed
water-driven wheels. I smelled imagination
among bestial hides by the gleam of fat,
seeking in all races a common ingenuity.
I envisaged an Africa flooded with such light

as alchemized the first fields of emmer wheat and barley,
when we savages dyed our pale dead with ochre,
and bordered our temples
with the ceremonial vulva of the conch
in the grey epoch of the obsidian adze.
I sowed the Sahara with rippling cereals,
my charity fertilized these aridities.

What was my field? Late sixteenth century.
My field was a dank acre. A Sussex don,
I taught the Jacobean anxieties: *The White Devil*.
Flamineo's torch startles the brooding yews.
The drawn end comes in strides. I loved my Duchess,
the white flame of her soul blown out between
the smoking cypresses. Then I saw children pounce
on green meat with a rat's ferocity.

I called them up and took the train to Bristol,
my blood the Severn's dregs and silver.
On Severn's estuary the pieces flash,
Iscariot's salary, patron saint of spies.
I thought, who cares how many million starve?
Their rising souls will lighten the world's weight
and level its gull-glittering waterline;
we left at sunset down the estuary.

England recedes. The forked white gull
screeches, circling back.
Even the birds are pulled back by their orbit,
even mercy has its magnetic field.
                        Back in the cabin,
I uncap the whisky, the porthole

mists with glaucoma. By the time I'm pissed,
England, England will be
that pale serrated indigo on the sea-line.
"You are so fortunate, you get to see the world—"
Indeed, indeed, sirs, I have seen the world.
Spray splashes the portholes and vision blurs.

Leaning on the hot rail, watching the hot sea,
I saw them far off, kneeling on hot sand
in the pious genuflections of the locust,
as Ponce's armoured knees crush Florida
to the funereal fragrance of white lilies.

    II

Now I have come to where the phantoms live,
I have no fear of phantoms, but of the real.
The Sabbath benedictions of the islands.
Treble clef of the snail on the scored leaf,
the Tantum Ergo of black choristers
soars through the organ pipes of coconuts.
Across the dirty beach surpliced with lace,
they pass a brown lagoon behind the priest,
pale and unshaven in his frayed soutane,
into the concrete church at Canaries;
as Albert Schweitzer moves to the harmonium
of morning, and to the pluming chimneys,
the groundswell lifts *Lebensraum, Lebensraum.*

Black faces sprinkled with continual dew—
dew on the speckled croton, dew

on the hard leaf of the knotted plum tree,
dew on the elephant ears of the dasheen.
Through Kurtz's teeth, white skull in elephant grass,
the imperial fiction sings. Sunday
wrinkles downriver from the Heart of Darkness.
The heart of darkness is not Africa.
The heart of darkness is the core of fire
in the white center of the holocaust.
The heart of darkness is the rubber claw
selecting a scalpel in antiseptic light,
the hills of children's shoes outside the chimneys,
the tinkling nickel instruments on the white altar;
Jacob, in his last card, sent me these verses:
"Think of a God who doesn't lose His sleep
if trees burst into tears or glaciers weep.
So, aping His indifference, I write now,
not Anno Domini: After Dachau."

III

The night maid brings a lamp and draws the blinds.
I stay out on the verandah with the stars.
Breakfast congealed to supper on its plate.

There is no sea as restless as my mind.
The promontories snore. They snore like whales.
Cetus, the whale, was Christ.
The ember dies, the sky smokes like an ash heap.
Reeds wash their hands of guilt and the lagoon
is stained. Louder, since it rained,
a gauze of sand flies hisses from the marsh.

Since God is dead, and these are not His stars,
but man-lit, sulphurous, sanctuary lamps,
it's in the heart of darkness of this earth
that backward tribes keep vigil of His Body,
in deya, lampion, and this bedside lamp.
Keep the news from their blissful ignorance.
Like lice, like lice, the hungry of this earth
swarm to the tree of life. If those who starve
like these rain-flies who shed glazed wings in light
grew from sharp shoulder blades their brittle vans
and soared towards that tree, how it would seethe—
ah, Justice! But fires
drench them like vermin, quotas
prevent them, and they remain
compassionate fodder for the travel book,
its paragraphs like windows from a train,
for everywhere that earth shows its rib cage
and the moon goggles with the eyes of children,
we turn away to read. Rimbaud learned that.
                                           Rimbaud, at dusk,
idling his wrist in water past temples
the plumed dates still protect in Roman file,
knew that we cared less for one human face
than for the scrolls in Alexandria's ashes,
that the bright water could not dye his hand
any more than poetry. The dhow's silhouette
moved through the blinding coinage of the river
that, endlessly, until we pay one debt,
shrouds, every night, an ordinary secret.

IV

The drawn sword comes in strides.
It stretches for the length of the empty beach;
the fishermen's huts shut their eyes tight.
A frisson shakes the palm trees,
and sweats on the traveller's tree.
They've found out my sanctuary. Philippe, last night:
"It had two gentlemen in the village yesterday, sir,
asking for you while you was in town.
I tell them you was in town. They send to tell you,
there is no hurry. They will be coming back."

In loaves of cloud, *and have not charity*,
the weevil will make a sahara of Kansas,
the ant shall eat Russia.
Their soft teeth shall make, *and have not charity*,
the harvest's desolation,
and the brown globe crack like a begging bowl,
and though you fire oceans of surplus grain,
and have not charity,

still, through thin stalks,
the smoking stubble, stalks
grasshopper: third horseman,
the leather-helmed locust.

The Season of Phantasmal Peace

Then all the nations of birds lifted together
the huge net of the shadows of this earth
in multitudinous dialects, twittering tongues,
stitching and crossing it. They lifted up
the shadows of long pines down trackless slopes,
the shadows of glass-faced towers down evening streets,
the shadow of a frail plant on a city sill—
the net rising soundless as night, the birds' cries soundless, until
there was no longer dusk, or season, decline, or weather,
only this passage of phantasmal light
that not the narrowest shadow dared to sever.

And men could not see, looking up, what the wild geese drew,
what the ospreys trailed behind them in silvery ropes
that flashed in the icy sunlight; they could not hear
battalions of starlings waging peaceful cries,
bearing the net higher, covering this world
like the vines of an orchard, or a mother drawing
the trembling gauze over the trembling eyes
of a child fluttering to sleep;
                              it was the light
that you will see at evening on the side of a hill
in yellow October, and no one hearing knew
what change had brought into the raven's cawing,
the killdeer's screech, the ember-circling chough

such an immense, soundless, and high concern
for the fields and cities where the birds belong,
except it was their seasonal passing, Love,
made seasonless, or, from the high privilege of their birth,
something brighter than pity for the wingless ones
below them who shared dark holes in windows and in houses,
and higher they lifted the net with soundless voices
above all change, betrayals of falling suns,
and this season lasted one moment, like the pause
between dusk and darkness, between fury and peace,
but, for such as our earth is now, it lasted long.

From

MIDSUMMER

[1984]

II

Companion in Rome, whom Rome makes as old as Rome,
old as that peeling fresco whose flaking paint
is the clouds, you are crouched in some ancient pensione
where the only new thing is paper, like young St. Jerome
with his rock vault. Tonsured, you're muttering a line
that your exiled country will soon learn by heart,
to a flaking, sunlit ledge where a pigeon gurgles.
Midsummer's furnace casts everything in bronze.
Traffic flows in slow coils, like the doors of a baptistry,
and even the kitten's eyes blaze with Byzantine icons.
That old woman in black, unwrinkling your sheet with a palm,
her home is Rome, its history is her house.
Every Caesar's life has shrunk to a candle's column
in her saucer. Salt cleans their bloodstained togas.
She stacks up the popes like towels in cathedral drawers;
now in her stone kitchen, under the domes of onions,
she slices a light, as thick as cheese, into epochs.
Her kitchen wall flakes like an atlas where, once,
Ibi dracones was written, where unchristened cannibals
gnawed on the dry heads of coconuts as Ugolino did.
Hell's hearth is as cold as Pompeii's. We're punished by bells
as gentle as lilies. Luck to your Roman elegies
that the honey of time will riddle like those of Ovid.
Corals up to their windows in sand are my sacred domes,

gulls circling a seine are the pigeons of my St. Mark's,
silver legions of mackerel race through our catacombs.

III

At the Queen's Park Hotel, with its white, high-ceilinged rooms,
I reenter my first local mirror. A skidding roach
in the porcelain basin slides from its path to Parnassus.
Every word I have written took the wrong approach.
I cannot connect these lines with the lines in my face.
The child who died in me has left his print on
the tangled bed linen, and it was his small voice
that whispered from the gargling throat of the basin.
Out on the balcony I remember how morning was:
It was like a granite corner in Piero della Francesca's
"Resurrection," the cold, sleeping foot
prickling like the small palms up by the Hilton.
On the dewy Savannah, gently revolved by their grooms,
snorting, delicate-ankled racehorses exercise,
as delicate-ankled as brown smoke from the bakeries.
Sweat darkens their sides, and dew has frosted the skins
of the big American taxis parked all night on the street.
In black asphalt alleys marked by a ribbon of sunlight,
the closed faces of shacks are touched by that phrase in Traherne:
"The corn was orient and immortal wheat,"
and the canefields of Caroni. With all summer to burn,
a breeze strolls down to the docks, and the sea begins.

VI

Midsummer stretches beside me with its cat's yawn.
Trees with dust on their lips, cars melting down
in its furnace. Heat staggers the drifting mongrels.
The capitol has been repainted rose, the rails
round Woodford Square the colour of rusting blood.
Casa Rosada, the Argentinian mood,
croons from the balcony. Monotonous lurid bushes
brush the damp clouds with the ideograms of buzzards
over the Chinese groceries. The oven alleys stifle.
In Belmont, mournful tailors peer over old machines,
stitching June and July together seamlessly.
And one waits for midsummer lightning as the armed sentry
in boredom waits for the crack of a rifle.
But I feed on its dust, its ordinariness,
on the faith that fills its exiles with horror,
on the hills at dusk with their dusty orange lights,
even on the pilot light in the reeking harbour
that turns like a police car's. The terror
is local, at least. Like the magnolia's whorish whiff.
All night, the barks of a revolution crying wolf.
The moon shines like a lost button.
The yellow sodium lights on the wharf come on.
In streets, dishes clatter behind dim windows.
The night is companionable, the future as fierce as
tomorrow's sun everywhere. I can understand

Borges's blind love for Buenos Aires,
how a man feels the streets of a city swell in his hand.

VII

Our houses are one step from the gutter. Plastic curtains
or cheap prints hide what is dark behind windows—
the pedalled sewing machine, the photos, the paper rose
on its doily. The porch rail is lined with red tins.
A man's passing height is the same size as their doors,
and the doors themselves usually no wider than coffins,
sometimes have carved in their fretwork little half-moons.
The hills have no echoes. Not the echo of ruins.
Empty lots nod with their palanquins of green.
Any crack in the sidewalk was made by the primal fault
of the first map of the world, its boundaries and powers.
By a pile of red sand, of seeding, abandoned gravel
near a burnt-out lot, a fresh jungle unfurls its green
elephants' ears of wild yams and dasheen.
One step over the low wall, if you should care to,
recaptures a childhood whose vines fasten your foot.
And this is the lot of all wanderers, this is their fate,
that the more they wander, the more the world grows wide.
So, however far you have travelled, your
steps make more holes and the mesh is multiplied—
or why should you suddenly think of Tomas Venclova,
and why should I care about whatever they did to Heberto
when exiles must make their own maps, when this asphalt
takes you far from the action, past hedges of unaligned flowers?

XI

My double, tired of morning, closes the door
of the motel bathroom; then, wiping the steamed mirror,
refuses to acknowledge me staring back at him.
With the softest grunt, he stretches my throat for the function
of scraping it clean, his dispassionate care
like a barber's lathering a corpse—extreme unction.
The old ritual would have been as grim
if the small wisps that curled there in the basin
were not hairs but minuscular seraphim.
He clips our moustache with a snickering scissors,
then stops, reflecting, in midair. Certain sadnesses
are not immense, but fatal, like the sense of sin
while shaving. And empty cupboards where her dresses
shone. But why flushing a faucet, its vortex
swivelling with bits of hair, could make some men's
hands quietly put aside their razors,
and sense their veins as filth floating downriver
after the dolorous industries of sex,
is a question swans may raise with their white necks,
that the cockerel answers quickly, treading his hens.

XIV

With the frenzy of an old snake shedding its skin,
the speckled road, scored with ruts, smelling of mould,
twisted on itself and reentered the forest
where the dasheen leaves thicken and folk stories begin.
Sunset would threaten us as we climbed closer
to her house up the asphalt hill road, whose yam vines
wrangled over gutters with the dark reek of moss,
the shutters closing like the eyelids of that mimosa
called Ti-Marie; then—lucent as paper lanterns,
lamplight glowed through the ribs, house after house—
there was her own lamp at the black twist of the path.
There's childhood, and there's childhood's aftermath.
She began to remember at the minute of the fireflies,
to the sound of pipe water banging in kerosene tins,
stories she told to my brother and myself.
Her leaves were the libraries of the Caribbean.
The luck that was ours, those fragrant origins!
Her head was magnificent, Sidone. In the gully of her voice
shadows stood up and walked, her voice travels my shelves.
She was the lamplight in the stare of two mesmerized boys
still joined in one shadow, indivisible twins.

XV

I can sense it coming from far, too, Maman, the tide
since day has passed its turn, but I still note
that as a white gull flashes over the sea, its underside
catches the green, and I promise to use it later.
The imagination no longer goes as far as the horizon,
but it keeps coming back. At the edge of the water
it returns clean, scoured things that, like rubbish,
the sea has whitened, chaste. Disparate scenes.
The pink and blue chattel houses in the Virgins
in the trade winds. My name caught in
the kernel of my great-aunt's throat.
A yard, an old brown man with a moustache
like a general's, a boy drawing castor-oil leaves in
great detail, hoping to be another Albrecht Dürer.
I have cherished these better than coherence
as the same tide for us both, Maman, comes nearer—
the vine leaves medalling an old wire fence
and, in the shade-freckled yard, an old man like a colonel
under the green cannonballs of a calabash.

XVIII

In the other 'eighties, a hundred midsummers gone
like the light of domestic paradise, the hedonist's
idea of heaven was a French kitchen's sideboard,
apples and clay carafes from Chardin to the Impressionists;
art was *une tranche de vie*, cheese or home-baked bread—
light, in their view, was the best that time offered.
The eye was the only truth, and whatever traverses
the retina fades when it darkens; the depth of *nature morte*
was that death itself is only another surface
like the canvas, since painting cannot capture thought.
A hundred midsummers gone, with the rippling accordion,
bustled skirts, boating parties, zinc-white strokes on water,
girls whose flushed cheeks wouldn't outlast their roses.
Then, like dried-up tubes, the coiled soldiers
piled up on the Somme, and Verdun. And the dead
less real than a spray burst of chrysanthemums,
the identical carmine for still life and for the slaughter
of youth. They were right—everything becomes
its idea to the painter with easel rifled on his shoulders.

XIX

Gauguin/i

On the quays of Papeete, the dawdling white-ducked colonists
drinking with whores whose skin is the copper of pennies
pretend, watching the wild skins of the light and shade,
that a straight vermouth re-creates the metropolis,
but the sun has scorched those memories from my head—
Cézanne bricking in color, each brick no bigger than a square inch,
the pointillists' dots like a million irises.
I saw in my own cheekbones the mule's head of a Breton,
the placid, implacable strategy of the Mongol,
the moustache like the downturned horns of a helmet;
the chain of my blood pulled me to darker nations,
though I looked like any other sallow, crumpled colon
stepping up to the pier that day from the customs launch.
I am Watteau's wild oats, his illegitimate heir.
Get off your arses, you clerks, and find your fate,
the devil's prayer book is the hymn of patience,
grumbling in the fog. Pack, leave! I left too late.

ii

I have never pretended that summer was paradise,
or that these virgins were virginal; on their wooden trays
are the fruits of my knowledge, radiant with disease,
and they offer you this, in their ripe sea-almond eyes,

their clay breasts glowing like ingots in a furnace.
No, what I have plated in amber is not an ideal, as
Puvis de Chavannes desired it, but corrupt—
the spot on the ginger lily's vulva, the plantain's phalloi,
the volcano that chafes like a chancre, the lava's smoke
that climbs to the sibilant goddess with its hiss.
I have baked the gold of their bodies in that alloy;
tell the Evangelists paradise smells of sulphur,
that I have felt the beads in my blood erupt
as my brush stroked their backs, the cervix
of a defrocked Jesuit numbering his chaplet.
I placed a blue death mask there in my Book of Hours
that those who dream of an earthly paradise may read it
as men. My frescoes in sackcloth to the goddess Maya.
The mangoes redden like coals in a barbecue pit,
patient as the palms of Atlas, the papaya.

XX

Watteau

The amber spray of trees feather-brushed with the dusk,
the ruined cavity of some spectral château, the groin
of a leering satyr eaten with ivy. In the distance, the grain
of some unreapable, alchemical harvest, the hollow at
the heart of all embarkations. Nothing stays green
in that prodigious urging towards twilight;
in all of his journeys the pilgrims are in fever
from the tremulous strokes of malaria's laureate.
So where is Cythera? It, too, is far and feverish,
it dilates on the horizon of his near-delirium, near
and then further, it can break like the spidery rigging
of his ribboned barquentines, it is as much nowhere
as these broad-leafed islands, it is the disease
of elephantine vegetation in Baudelaire,
the tropic bug in the Paris fog. For him, it is the mirror
of what is. Paradise is life repeated spectrally,
an empty chair echoing the emptiness.

XXI

A long, white, summer cloud, like a cleared linen table,
makes heaven emptier, like after-dinner Sundays
when the Bible begs to be lifted, and the old terrifying verses
raise a sandstorm and bone-white Palestinian rocks
where a ram totters for purchase, bleating like Isaiah.
Dry rage of the desert fathers that scared a child,
the Baptist crying by the cracked river basin, curses
that made the rose an intellectual fire.
Through the skull's stone eyes, the radiant logwood
consumes this August, and a white sun sucks
sweat from the desert. A shadow marks the Word.
I have forgotten a child's hope of the resurrection,
bodies locked up in musting cupboard drawers
among the fish knives and the napery (all the dead earth holds),
to be pulled open at the hour of our birth—
the cloud waits in emptiness for the apostles,
for the fruit, wine amphoras, mutton on groaning trestles,
but only the servant knows heaven is still possible,
some freckled Martha, radiant, dependable,
singing a hymn from your childhood while she folds
her Saviour like a white napkin in the earth.

XXIII

With the stampeding hiss and scurry of green lemmings,
midsummer's leaves race to extinction like the roar
of a Brixton riot tunnelled by water hoses;
they seethe towards autumn's fire—it is in their nature,
being men as well as leaves, to die for the sun.
The leaf stems tug at their chains, the branches bending
like Boer cattle under Tory whips that drag every wagon
nearer to apartheid. And, for me, that closes
the child's fairy tale of an antic England—fairy rings,
thatched cottages fenced with dog roses,
a green gale lifting the hair of Warwickshire.
I was there to add some colour to the British theatre.
"But the blacks can't do Shakespeare, they have no experience."
This was true. Their thick skulls bled with rancour
when the riot police and the skinheads exchanged quips
you could trace to the Sonnets, or the Moor's eclipse.
Praise had bled my lines white of any more anger,
and snow had inducted me into white fellowships,
while Calibans howled down the barred streets of an empire
that began with Caedmon's raceless dew, and is ending
in the alleys of Brixton, burning like Turner's ships.

XXV

The sun has fired my face to terra-cotta.
It carries the heat from his kiln all through the house.
But I cherish its wrinkles as much as those on blue water.
Gnats drill little holes around a saw-toothed cactus,
a furnace has curled the knives of the oleander,
and a branch of the logwood blurs with wild characters.
A stone house waits on the steps. Its white porch blazes.
I tell you a promise brought to me by the surf:
You shall see transparent Helen pass like a candle
flame in sunlight, weightless as woodsmoke that hazes
the sand with no shadow. My palms have been sliced by the twine
of the craft I have pulled at for more than forty years.
My Ionia is the smell of burnt grass, the scorched handle
of a cistern in August squeaking to rusty islands;
the lines I love have all their knots left in.
Through the stunned afternoon, when it's too hot to think
and the muse of this inland ocean still waits for a name,
and from the salt, dark room, the tight horizon line
catches nothing, I wait. Chairs sweat. Paper crumples the floor.
A lizard gasps on the wall. The sea glares like zinc.
Then, in the door light: not Nike loosening her sandal,
but a girl slapping sand from her foot, one hand on the frame.

XXVI

Before that thundercloud breaks from its hawsers,
those ropes of rain, a wind makes the sea grapes wince,
and the reef signals its last flash of lime.
Feeling her skin cool, the housemaid August
runs into the yard to pull down clouds, like a laundress,
from the year's meridian, her mouth stuffed with wooden pins.
She's seen these flashes of quartz, she knows it's time
for the guests on the beach to come up to the house,
and, hosing sand from scorched feet, let the hinges rust
in holes for another year. But an iron band
still binds their foreheads: the bathers stand
begging the dark clouds, whose spinnakers race over the dunes,
for one more day. Here, the salt vine dries
as fast as it grows, and before you look, a year's gone
with your shadow. The temperate homilies can't
take root in sand; the cicada can fiddle his tunes
all year, if he likes, to the twig-brown ant.
The cloud passes high like a god staying his powers—
the pocked sand dries, umbrellas reopen like flowers—
but those who measure midsummer by a year's trials
have felt a chill grip an ankle. They put down their books
to count the children crouched over pools, and the idolaters
angling themselves to the god's face, like sundials.

XXVII

Certain things here are quietly American—
that chain-link fence dividing the absent roars
of the beach from the empty ball park, its holes
muttering the word umpire instead of empire;
the grey, metal light where an early pelican
coasts, with its engine off, over the pink fire
of a sea whose surface is as cold as Maine's.
The light warms up the sides of white, eager Cessnas
parked at the airstrip under the freckling hills
of St. Thomas. The sheds, the brown, functional hangar,
are like those of the Occupation in the last war.
The night left a rank smell under the casuarinas,
the villas have fenced-off beaches where the natives walk,
illegal immigrants from unlucky islands
who envy the smallest polyp its right to work.
Here the wetback crab and the mollusc are citizens,
and the leaves have green cards. Bulldozers jerk
and gouge out a hill, but we all know that the dust
is industrial and must be suffered. Soon—
the sea's corrugations are sheets of zinc
soldered by the sun's steady acetylene. This
drizzle that falls now is American rain,
stitching stars in the sand. My own corpuscles
are changing as fast. I fear what the migrant envies:

the starry pattern they make—the flag on the post office—
the quality of the dirt, the fealty changing under my foot.

XXVIII

Something primal in our spine makes the child swing
from the gnarled trapeze of a sea-almond branch.
I have been comparing the sea-almond's shapes to the suffering
in Van Gogh's orchards. And that, too, is primal. A bunch
of sea grapes hangs over the calm sea. The shadows
I shovel with a dry leaf are as warm as ash, as
noon jerks towards its rigid, inert centre.
Sunbathers broil on their grid, the shallows they enter
are so warm that out in the reef the blear grouper lunges
at nothing, teased by self-scaring minnows.
Abruptly remembering its job, a breaker glazes
the sand that dries fast. For hours, without a heave,
the sea suspires through the deep lungs of sponges.
In the thatched beach bar, a clock tests its stiff elbow
every minute and, outside, an even older iguana
climbs hand over claw, as unloved as Quasimodo,
into his belfry of shade, swaying there. When a
cloud darkens, my terror caused it. Lizzie and Anna
lie idling on different rafts, their shadows under them.
The curled swell has the clarity of lime.
In two more days my daughters will go home.
The frame of human happiness is time,
the child's swing slackens to a metronome.
Happiness sparkles on the sea like soda.

XXX

Gold dung and urinous straw from the horse garages,
click-clop of hooves sparking cold cobblestone.
From bricked-in carriage yards, exhaling arches
send the stale air of transcendental Boston—
tasselled black hansoms trotting under elms,
tilting their crops to the shade of Henry James.
I return to the city of my exile down Storrow Drive,
the tunnel with its split seraphs flying *en face*,
with finite sorrow; blocks long as paragraphs
pass in a style to which I'm not accustomed,
since, if I were, I would have been costumed
to drape the cloaks of couples who arrive
for dinner, drawing their chairs from tables where each glass,
catching the transcendental clustered lights,
twirled with perceptions. Style is character—
so my forehead crusts like brick, my sockets char
like a burnt brownstone in the Negro Quarter;
but when a fog obscures the Boston Common
and, up Beacon Hill, the old gas standards stutter
to save their period, I see a black coachman,
with gloves as white as his white-ankled horse,
who counts their laughter, their lamplit good nights,
then jerks the reins of his brass-handled hearse.

XXXIII

[*for Robert Fitzgerald*]

Those grooves in that forehead of sand-coloured flesh
were cut by declining keels, and the crow's foot
that prints an asterisk by unburied men
reminds him how many more by the Scamander's
gravel fell and lie waiting for their second fate.
Who next should pull his sword free of its mesh
of weeds and hammer at the shield
of language till the wound and the word fit?
A whole war is fought backwards to its cause.
Last night, the Trojan and the Greek commanders
stood up like dogs when his strange-smelling shadow
hung loitering round their tents. Now, at sunrise,
the dead begin to cough, each crabwise hand
feels for its lance, and grips it like his pen.
A helmsman drowns in an inkblot, an old man wanders
a pine-gripped islet where his wound was made.
Entering a door-huge dictionary, he finds that clause
that stopped the war yesterday; his pulse starts the gavel
of hexametrical time, the V's of each lifted blade
pull from Connecticut, like the hammers of a piano
without the sound, as the wake, reaching gravel,
recites in American: *"Arma virumque cano..."*

XXXV

Mud. Clods. The sucking heel of the rain-flinger.
Sometimes the gusts of rain veered like the sails
of dragon-beaked vessels dipping to Avalon
and mist. For hours, driving along
the skittering ridges of Wales, we carried the figure
of Langland's Ploughman on the rain-seeded glass,
matching the tires with his striding heels,
while splintered puddles dripped from the roadside grass.
Once, in the drizzle, a crouched, clay-covered ghost
rose in his pivot, and the turning disk of the fields
with their ploughed stanzas sang of a freshness lost.
Villages began. We had crossed into England—
the fields, not their names, were the same. We found a caff,
parked in a thin drizzle, then crammed into a pew
of red leatherette. Outside, with thumb and finger,
a careful sun was picking the lint from things.
The sun brightened like a sign, the world was new
while the cairns, the castled hillocks, the stony kings
were scabbarded in sleep, yet what made me think
that the crash of chivalry in a kitchen sink
was my own dispossession? I could sense, from calf
to flinging wrist, my veins ache in a knot.
There was mist on the window. I rubbed it and looked out
at the helmets of wet cars in the parking lot.

XXXVI

The oak inns creak in their joints as light declines
from the ale-coloured skies of Warwickshire.
Autumn has blown the froth from the foaming orchards,
so white-haired regulars draw chairs nearer the grate
to spit on logs that crackle into leaves of fire.
But they grow deafer, not sure if what they hear
is the drone of the abbeys from matins to compline,
or the hornet's nest of a chain saw working late
on the knoll up there back of the Norman chapel.
Evening loosens the moth, the owl shifts its weight,
a fish-mouthed moon swims up from wavering elms,
but four old men are out on the garden benches,
talking of the bows they have drawn, their strings of wenches,
their coined eyes shrewdly glittering like the Thames'
estuaries. I heard their old talk carried
through cables laid across the Atlantic bed,
their gossip rustles like an apple orchard's
in my own head, and I can drop their names
like familiars—those bastard grandsires
whose maker granted them a primal pardon—
because the worm that cores the rotting apple
of the world and the hornet's chain saw cannot touch the words
of Shallow or Silence in their fading garden.

XXXIX

The grey English road hissed emptily under the tires
since the woods still drizzled. The sound was like foam
mixed with island rain, but the rain was Berkshire's.
He said a white hare would startle itself like a tuft
on the road's bare scalp. But, wherever it came from,
the old word "hare" shivered like "weald" or "croft"
or the peeled white trunk with a wound in "atheling."
I hated fables. The wheezing beeches were fables,
and the wild, wet mustard. As for the mist, gathering
from the mulch of black leaves in which the hare hid
in clenched concentration—muttering prayers, bead-eyed,
haunch-deep in nettles—the sooner it disappeared
the better. Something branched in that countryside
losing ground to the mist, its old roads brown as blood.
The white hare had all of England on which to brood
with its curled paws—from the age of skins and woad,
from Saxon settlements fenced with stakes, and thick
fires of peat smoke, down to thin country traffic.
He turned on the fog lights. It was on this road,
on this ridge of earth long since swept bare
of his mud prints, that my bastard ancestor swayed
transfixed by the trembling, trembling thing that stood
its ground, ears pronged, nibbling him into a hare.

XLI

The camps hold their distance—brown chestnuts and grey smoke
that coils like barbed wire. The profit in guilt continues.
Brown pigeons goose-step, squirrels pile up acorns like little shoes,
and moss, voiceless as smoke, hushes the peeled bodies
like abandoned kindling. In the clear pools, fat
trout rising to lures bubble in umlauts.
Forty years gone, in my island childhood, I felt that
the gift of poetry had made me one of the chosen,
that all experience was kindling to the fire of the Muse.
Now I see her in autumn on that pine bench where she sits,
their nut-brown ideal, in gold plaits and *lederhosen*,
the blood drops of poppies embroidered on her white bodice,
the spirit of autumn to every Hans and Fritz
whose gaze raked the stubble fields when the smoky cries
of rooks were nearly human. They placed their cause in
her cornsilk crown, her cornflower iris,
winnower of chaff for whom the swastikas flash
in skeletal harvests. But had I known then
that the fronds of my island were harrows, its sand the ash
of the distant camps, would I have broken my pen
because this century's pastorals were being written
by the chimneys of Dachau, of Auschwitz, of Sachsenhausen?

XLII

Chicago's avenues, as white as Poland.
A blizzard of heavenly coke hushes the ghettos.
The scratched sky flickers like a TV set.
Down Michigan Avenue, slow as the glacial prose
of historians, my taxi crawls. The stalled cars are as frozen
as the faces of cloaked queues on a Warsaw street,
or the hands of black derelicts flexing over a fire-
barrel under the El; above, the punctured sky
is needled by rockets that keep both Empires high.
It will be both ice and fire. In the sibyl's crystal
the globe is shaken with ash, with a child's *frisson*.
It'll be like this. A bird cry will sound like a pistol
down the avenues. Cars like dead horses, their muzzles
foaming with ice. From the cab's dashboard, a tinny
dispatcher's voice warns of more snow. A picture
lights up the set—first, indecipherable puzzles;
then, in plain black and white, a snow slope with pines
as shaggy as the manes of barbarian ponies;
then, a Mongol in yak's skin, teeth broken as dice,
grinning at the needles of the silent cities
of the plains below him up in the Himalayas,
who slaps the snow from his sides and turns away as,
in lance-like birches, the horde's ponies whinny.

XLIII

Tropic Zone/i

A white dory, face down, its rusted keel staining
the hull, bleeds under the dawn leaves of an almond.
Vines grip the seawall and drop like olive-green infantry
over from Cuba. This is my ocean, but it is speaking
another language, since its accent changes around
different islands. The wind is up early, campaigning
with the leaflets of seagulls, but from the balcony
of the guesthouse, I resist the return
of this brightening noun whose lines must be translated
into *"el mar"* or *"la mar,"* and death itself to *"la muerte."*
A rusty sparrow alights on a rustier rain gauge
in the front garden, but every squeak addresses
me in testy Spanish. "Change to a light shirt. A
walk on our beach should teach you our S's
as the surf says them. You'll recognize hovels,
rotting fishnets. Also why a white dory was shot
for being a gringo." I go back upstairs,
for so much here is the Empire envied and hated
that whether one chooses to say *"ven-thes"* or *"ven-ces"*
involves the class struggle as well. So, be discreet.
Changed to a light shirt, I walk out to Cervantes Street.
Shadow-barred. A water sprinkler or a tank approaches.
The corners are empty. The boulevards open like novels
waiting to be written. Clouds like the beginnings of stories.

ii

The sun is wholly up now; things are white or green:
clouds, hills, walls, leaves on the walls, and their shadows;
dew turns into dust on the quiet municipal cedars.
The sprinkler rolls past as "the wrong done to our fathers"
weeps along empty streets, down serene *avenidas*
named after stone poets, but the sprinkling only grows
traffic. When noon strikes the present-arms pose
of sentries in boxes before the Palace of Governors,
history will pierce your memory like a migraine;
but however their flame trees catch, the green winds smell
    lime-scented,
the indigo hills lie anchored in seas of cane
as deep as my island's, I know I would feel disoriented
in Oriente, my tongue dried to a coral stone.
Along white-walled, palm-splashed Condado, the breeze smells
of a dialect so strong it is not disinfected
by the exhausts of limousines idling outside the hotels,
while, far out, unheard, the grinding reef of the Morro
spits out like corals the indigestible sorrow
of the Indian, bits for the National Museum.
Blue skies convert all genocide into fiction,
but a man, drawn to the seawall, crouches like a question
or a prayer, and my own prayer is to write
lines as mindless as the ocean's of linear time,
since time is the first province of Caesar's jurisdiction.

iii

Above hot tin billboards, above Hostería del Mar,
wherever the Empire has raised the standard of living

by blinding high rises, gestures are made to the culture
of a remorseful past, whose artists must stay unforgiving
even when commissioned. If the white architectural mode is
International Modern, the décor must be the Creole's,
so, in a terra-cotta lobby with palms, a local jingle
gurgles of a new *cerveza*, frost-crusted and golden,
right next to a mural that has nationalized Eden
in vehement acrylics, and this universal theme
sees the golden beer, the gold mines, "the gold of their bodies"
as one, and our two tropics as erogenous zones.
A necklace of emerald islands is fringed with lace
starched as the ruffles of Isabella's bodice,
now the white-breasted Niña and Pinta and Santa María
bring the phalli of lances penetrating a jungle
whose vines spread apart to a parrot's primal scream.
Then, shy as the ferns their hands are bending, stare
fig-nippled maidens with faces calm as stones,
and, as is the case with so many revolutions,
the visitor doubts the murals and trusts the beer.

iv

Noon empties balconies, but the arched eyebrows
of the plaza are not amazed at the continuum—
a fly drilling holes in a snoring peon's face,
the arched shade of patios humming with audible heat,
and long-fingered shadows retracting to a fist.
The statue's sword arm is tired, he'd like to dismount
from his leaf-green stallion and curl up in the shade
with the rest of his country. And that's how it was
in the old scenarios, a backdrop for the hectic
conscience of the gringo with his Wasp's rage at tedium,

but now in the banana republics, whose bunches of recruits
look green in fatigues, techniques of camouflage
have taught the skill of slitting stomachs like fruits,
and a red star without a sickle is stitched to a flag.
Now the women who were folded over wrought-iron
balconies like bedsteads, their black manes hanging down,
are not whores with roses but dolls broken in half.
On a wall a bleeding VIVA! hieroglyphs speeches
that lasted four hours in marathon dialectic.
Sand-coloured mongrels prowl round a young Antigone,
her face flat as an axe of pre-Columbian stone.
At the movies, I still love it when gap-toothed bandidos laugh
in growling pidgin, then grin at the sudden contradiction
of roses stitching their guts. In colonial fiction
evil remains comic and only achieves importance
when the gringo crosses the plaza, flayed by the shadows of fronds.

v

"Wherever a thought can go back seventy years
there is hope for tradition in these tropical zones."
The old men mutter in white suits, elbows twitching like pigeons
on their canes, under the dusty leaves of the almonds
that grant them asylum from paths ruined by bicycles,
from machines with umbrellas dispensing franks and cones.
Their revolution is that things come in circles.
The socialists do not appreciate that.
But old almonds do, and there is appreciation
in the tilt of a cannon's chin to the horizon,
and applause from the seawall when a crash of lace
is like that moment of flamenco, *Ah, mi corazón,*
that moment of flamenco when the dancer's

heels rattled like gunfire and, above her tilted comb, her
clapping hands were like midnight on a clock!
For each old man, in his white panama hat,
there is no ideology in the light: this one
shakes his cane like a question without answers,
that one riddles the militia with his smiles,
another one leans backwards in a coma
of silence—when lilies opened like Victrola horns,
when dusk spread feathers like a fighting cock,
and down the Sunday promenade for miles
the Civil Guard kept playing "La Paloma"
and gulls, like doves, waltzed to the gusting lace
and everyone wore white and there was grace.

   vi

You've forgotten the heat. It could burn from a zinc fence.
Not even the palms on the seafront quietly stir.
The Empire sneers at all thoughts in the future tense.
Only the shallows of this inland ocean mutter
lines from another sea, which this one resembles—
myths of analogous islands of olive and myrtle,
the dream of the drowsing Gulf. Although her temples,
white blocks against green, are hotels, and her stoas
shopping malls, in time they will make good ruins;
so what if the hand of the Empire is as slow as
a turtle signing the surf when it comes to treaties?
Genius will come to contradict history,
and that's there in their brown bodies, in the olives of eyes,
as when the pimps of demotic Athens threaded the chaos
of Asia, and girls from the stick villages, henna-whores,
were the hetaerae. The afternoon tide ebbs, and the stench

of further empires—rising from berries that fringe
the hems of tyrants and beaches—reaches a bench
where clouds descend their steps like senates passing,
no different from when, under leaves of rattling myrtle,
they shared one shade, the poet and the assassin.

vii

Imagine, where sand is now, the crawling lava
of military concrete. Sprinkle every avenue with the grey
tears of the people's will. Tyranny brings over
its colonies this disorientation of weather. A new ogre
erects his bronzes over the parks, though the senate
of swallows still arranges itself on benches
for the usual agenda, and three men can still argue
under a changed street sign, but the streets are emptier
and the mouth dry. Imagine the fading hysteria
of peeling advertisements, and note how all the graffiti agree
with the government. You might say, Yes, but here are mountains,
park benches, working fountains, a brass band on Sundays,
here the baker still gives a special twist to the end
of his father's craft, until one morning you notice
that the three men talk softly, that mothers call
from identical windows for their children to come home,
that the smallest pamphlet is stamped with a single star.
The days feel longer, people resemble their cars
that are gray as their uniforms. In the millennium,
most men, at night, sleep with their eyes to the wall.

viii

If you were here, in this white room, in this hotel
whose hinges stay hot, even in the wind off the sea,
you would sprawl, knocked out by *la hora de siesta;*
you couldn't rise for the resurrection bell,
or the sea's gong ringing with silver, you'd stay down.
If you were touched, you'd only change that gesture
to a runner's in that somnambulist's marathon.
And I'd let you sleep. Things topple gradually
when the alarm clock, with its conductor's baton,
begins at one: the cattle fold their knees;
in the quiet pastures, only a mare's tail switches,
feather-dusting flies, drunk melons roll into ditches,
and gnats keep spiralling to their paradise.
Now the first gardener, under the tree of knowledge,
forgets that he's Adam. In the ribbed air
each patch of shade dilates like an oasis
to the tired butterfly, a green lagoon for anchor.
Down the white beach, calm as a forehead
that has felt the wind, a sacramental stasis
would bring you sleep, which is midsummer's crown,
sleep that divides its lovers without rancour,
sweat without sin, the furnace without fire,
calm without self, the dying with no fear,
as afternoon removes those window bars
that striped your sleep like a kitten's, or a prisoner's.

XLIX

A wind-scraped headland, a sludgy, dishwater sea,
another storm-darkened village with fences of crucified tin.
Give it up to a goat in the rain, whose iron muzzle
can take anything, or to those hopping buzzards
trailing their torn umbrellas in a silvery drizzle
that slimes everything; on the horizon,
the sea's silver language shines like another era,
and, seasick of poverty, my mind is out there.
A storm has wrecked the island, the beach is a mess,
a bent man, crouching, crosses it, cuffed by the wind;
from that gap of blue, with seraphic highmindedness,
the frigate birds are crying that foul weather lifts the soul,
that the sodden red rag of the heart, when it has dried,
will flutter like a lifeguard's flag from its rusty pole.
Though I curse the recurrence of each shining omen,
the sun will come out, and warm up my right hand
like that old crab flexing its fingers outside its hole.
Frail from damp holes, the courageous, pale bestiary
of the sand seethes, the goat nuzzles, head bent
among flashing tins, and the light's flood tide
stutters up to a sandbar in the estuary,
where, making the most of its Egyptian moment,
the heron halts its abrupt, exalted stride—
then a slow frieze of sunlit pelicans.

L

I once gave my daughters, separately, two conch shells
that were dived from the reef, or sold on the beach, I forget.
They use them as doorstops or bookends, but their wet
pink palates are the soundless singing of angels.
I once wrote a poem called "The Yellow Cemetery,"
when I was nineteen. Lizzie's age. I'm fifty-three.
These poems I heaved aren't linked to any tradition
like a mossed cairn; each goes down like a stone
to the seabed, settling, but let them, with luck, lie
where stones are deep, in the sea's memory.
Let them be, in water, as my father, who did watercolours,
entered his work. He became one of his shadows,
wavering and faint in the midsummer sunlight.
His name was Warwick Walcott. I sometimes believe
that his father, in love or bitter benediction,
named him for Warwickshire. Ironies
are moving. Now, when I rewrite a line,
or sketch on the fast-drying paper the coconut fronds
that he did so faintly, my daughters' hands move in mine.
Conches move over the sea-floor. I used to move
my father's grave from the blackened Anglican headstones
in Castries to where I could love both at once—
the sea and his absence. Youth is stronger than fiction.

LI

Since all of your work was really an effort to appease
the past, a need to be admitted among your peers,
let the inheritors question the sibyl and the Sphinx,
and learn that a raceless critic is a primate's dream.
You were distressed by your habitat, you shall not find peace
till you and your origins reconcile; your jaw must droop
and your knuckles scrape the ground of your native place.
Squat on a damp rock round which white lilies stiffen,
pricking their ears; count as the syllables drop
like dew from primeval ferns; note how the earth drinks
language as precious, depending upon the race.
Then, on dank ground, using a twig for a pen,
write Genesis and watch the Word begin.
Elephants will mill at their water hole to trumpet a
new style. Mongoose, arrested in rut,
and saucer-eyed mandrills, drinking from the leaves,
will nod as a dew-lapped lizard discourses on "Lives
of the Black Poets," gripping a branch like a lectern for better
delivery. Already, up in that simian Academe,
a chimp in bifocals, his lower lip a jut,
tears misting the lenses, is turning your *Oeuvres Complètes*.

LII

I heard them marching the leaf-wet roads of my head,
the sucked vowels of a syntax trampled to mud,
a division of dictions, one troop black, barefooted,
the other in redcoats bright as their sovereign's blood;
their feet scuffled like rain, the bare soles with the shod.
One fought for a queen, the other was chained in her service,
but both, in bitterness, travelled the same road.
Our occupation and the Army of Occupation
are born enemies, but what mortar can size
the broken stones of the barracks of Brimstone Hill
to the gaping brick of Belfast? Have we changed sides
to the moustached sergeants and the horsy gentry
because we serve English, like a two-headed sentry
guarding its borders? No language is neutral;
the green oak of English is a murmurous cathedral
where some took umbrage, some peace, but every shade, all,
helped widen its shadow. I used to haunt the arches
of the British barracks of Vigie. There were leaves there,
bright, rotting like revers or epaulettes, and the stenches
of history and piss. Leaves piled like the dropped aitches
of soldiers from rival shires, from the brimstone trenches
of Agincourt to the gas of the Somme. On Poppy Day
our schools bought red paper flowers. They were for Flanders.
I saw Hotspur cursing the smoke through which a popinjay
minced from the battle. Those raging commanders

from Thersites to Percy, their rant is our model.
I pinned the poppy to my blazer. It bled like a vowel.

LIII

There was one Syrian, with his bicycle, in our town.
I didn't know if he was a Syrian or an Assyrian.
When I asked him his race, about which Saroyan had written
that all that was left were seventy thousand Assyrians,
where were sixty-nine thousand nine hundred and ninety-nine?
he didn't answer, but smiled at the length of our street.
His pupils flashed like the hot spokes of a chariot,
or the silver wires of his secondhand machine.
I should have asked him about the patterns of birds
migrating in Aramaic, or the correct
pronunciation of wrinkled rivers like "Tagus."
Assyria was far as the ancient world that was taught us,
but then, so was he, from his hot-skinned camels and tents.
I was young and direct and my tense
was the present; if I, in my ignorance,
had distorted time, it was less than some tyrant's
indifference that altered his future.
He wore a white shirt. A black hat. His bicycle
had an iron basket in front. It moved through the mirage
of sugar-cane fields, crediting suits to the cutters.
Next, two more Syrians appeared. All three shared a store
behind which they slept. After that, there was
a sign with that name, so comical to us, of mythical
spade-bearded, anointed, and ringleted kings: ABDUL.
But to me there were still only seventy thousand

Assyrians, and all of them lived next door
in a hot dark room, muttering a language whose sound
had winged lions in it, and birds cut into a wall.

LIV

The midsummer sea, the hot pitch road, this grass, these shacks
    that made me,
jungle and razor grass shimmering by the roadside, the edge of art;
wood lice are humming in the sacred wood,
nothing can burn them out, they are in the blood;
their rose mouths, like cherubs, sing of the slow science
of dying—all heads, with, at each ear, a gauzy wing.
Up at Forest Reserve, before branches break into sea,
I looked through the moving, grassed window and thought
    "pines,"
or conifers of some sort. I thought, they must suffer
in this tropical heat with their child's idea of Russia.
Then suddenly, from their rotting logs, distracting signs
of the faith I betrayed, or the faith that betrayed me—
yellow butterflies rising on the road to Valencia
stuttering "yes" to the resurrection; "yes, yes is our answer,"
the gold-robed Nunc Dimittis of their certain choir.
Where's my child's hymnbook, the poems edged in gold leaf,
the heaven I worship with no faith in heaven,
as the Word turned toward poetry in its grief?
Ah, bread of life, that only love can leaven!
Ah, Joseph, though no man ever dies in his own country,
the grateful grass will grow thick from his heart.

Bibliography and Index

Books of Poetry by Derek Walcott

25 Poems. Port of Spain, Trinidad, B.W.I.: Guardian Commercial Printery, 1948.
Epitaph for the Young: XII Cantos. Barbados: Advocate, 1949.
Poems. Jamaica: City Printery, 1951.
In a Green Night: Poems 1948–1960. London: Jonathan Cape, 1962.
Selected Poems. New York: Farrar, Straus & Company, 1964. Part I comprises a selection from *In a Green Night* (Cape, 1962); Parts II and III, poems that were previously uncollected.
The Castaway and Other Poems. London: Jonathan Cape, 1965.
The Gulf and Other Poems. London: Jonathan Cape, 1969.
The Gulf. New York: Farrar, Straus and Giroux, 1970. This volume includes not only the whole of *The Gulf and Other Poems* (Cape, 1969) but selections from *The Castaway and Other Poems* (Cape, 1965).
Another Life. FSG, 1973; Cape, 1973. Second American paperback edition (no change in text of poem), with Introduction by Robert D. Hamner. Washington, D.C.: Three Continents Press, 1982.
Sea Grapes. Cape, 1976; FSG (slightly revised edition), 1976.
The Star-Apple Kingdom. FSG, 1979; Cape, 1980.
The Fortunate Traveller. FSG, 1981; Faber and Faber, 1982.
Midsummer. FSG, 1984; Faber, 1984.

A complete bibliography through 1984—*Derek Walcott: An Annotated Bibliography of His Works* by Irma Goldstraw—was brought out in that year by Garland Publishing (New York and London).

The poems in *Collected Poems* are taken from the most recently published volumes in which they appear.

Index of Titles

Adam's Song 302
Air 113
"A long, white, summer cloud, like a cleared linen table," XXI 482
Another Life 141
As John to Patmos 5
"At the Queen's Park Hotel, with its white, high-ceilinged rooms," III 471
"A wind-scraped headland, a sludgy, dishwater sea," XLIX 503

Banyan Tree, Old Year's Night, The 48
Beachhead 410
"Before that thundercloud breaks from its hawsers," XXVI 485
Bleecker Street, Summer 40
Blues 111
Brise Marine 43

Castaway, The 57
"Certain things here are quietly American—," XXVII 486
Che 123
"Chicago's avenues, as white as Poland." XLII 495
City's Death by Fire, A 6
Codicil 97
Cold Spring Harbor 131
"Companion in Rome, whom Rome makes as old as Rome," II 469
Coral 73
Crusoe's Island 68
Crusoe's Journal 92

Dark August 329

Early Pompeian 446
Easter 452
Egypt, Tobago 368

Elegy 109
Endings 326
Europa 418
Exile 100

Far Cry from Africa, A 17
Fist, The 327
Flock, The 77
Forest of Europe 375
Fortunate Traveller, The 456
From This Far 414

Gauguin 479
Glory Trumpeter, The 64
Goats and Monkeys 83
God Rest Ye Merry, Gentlemen 91
"Gold dung and urinous straw from the horse garages," XXX 489
Gulf, The 104
Guyana 115

Harbour, The 7
Homage to Edward Thomas 103
Homecoming: Anse La Raye 127
Hotel Normandie Pool, The 439
Hurucan 423

"I can sense it coming from far, too, Maman, the tide," XV 477
"I heard them marching the leaf-wet roads of my head," LII 506
In a Green Night 50
"In the other 'eighties, a hundred midsummers gone," XVIII 478
"I once gave my daughters, separately, two conch shells," L 504
Islands 52

Jean Rhys 427

Koenig of the River 379

Lampfall 95
Landfall, Grenada 125
Laventille 85
Lesson for This Sunday, A 38

Letter from Brooklyn, A *41*
Liberator, The *430*
Love after Love *328*
Love in the Valley *133*

Man Who Loved Islands, The *420*
Map of Europe, A *66*
Map of the New World *413*
Mass Man *99*
"Midsummer stretches beside me with its cat's yawn," VI *472*
Midsummer, Tobago *333*
Missing the Sea *63*
Morning Moon, The *338*
"Mud. Clods. The sucking heel of the rain-flinger," XXXV *491*
"My double, tired of morning, closes the door," XI *475*

Names *305*
Nearing Forty *136*
Negatives *124*
New World *300*
Nights in the Gardens of Port of Spain *67*
North and South *405*

Oddjob, a Bull Terrier *334*
Old New England *399*
Orient and Immortal Wheat *36*
Origins *11*
"Our houses are one step from the gutter. Plastic curtains," VII *474*

Parang *33*
Piano Practice *403*
Pocomania *31*
Polish Rider, The *47*
Prelude *3*
Preparing for Exile *304*

Return to D'Ennery; Rain *28*
Ruins of a Great House *19*

Sabbaths, W.I. *362*
Saddhu of Couva, The *372*

Sainte Lucie 309
Schooner *Flight*, The 345
Sea Canes 331
Sea-Chantey, A 44
Sea Grapes 297
Sea Is History, The 364
Season of Phantasmal Peace, The 464
"Since all of your work was really an effort to appease," LI 505
"Something primal in our spine makes the child swing," XXVIII 488
Spoiler's Return, The 432
Star 130
Star-Apple Kingdom, The 383
Sunday Lemons 298
Swamp, The 59

Tales of the Islands 22
Tarpon 61
"The camps hold their distance—brown chestnuts and grey smoke," XLI 494
"The grey English road hissed emptily under the tires," XXXIX 493
"The midsummer sea, the hot pitch road, this grass, these shacks that made me," LIV 510
"The oak inns creak in their joints as light declines," XXXVI 492
"There was one Syrian, with his bicycle, in our town," LIII 508
"The sun has fired my face to terra-cotta," XXV 484
"Those grooves in that forehead of sand-coloured flesh," XXXIII 490
To Return to the Trees 339
Tropic Zone 496
Two Poems on the Passing of an Empire 35

Upstate 401

Verandah 89
Village Life, A 79
Volcano 324

Wales 455
Walk, The 138
Watteau 481
Winding Up 336
"With the frenzy of an old snake shedding its skin," XIV 476
"With the stampeding hiss and scurry of green lemmings," XXIII 483